D1534918

MICHIGAN MOLECULAR INSTITUTE
1910 WEST ST. ANDREWS ROAD
MIDLAND, MICHIGAN 48640

PROPERTIES OF
ADVANCED
SEMICONDUCTOR
MATERIALS

PROPERTIES OF ADVANCED SEMICONDUCTOR MATERIALS

GaN, AlN, InN, BN, SiC, SiGe

Edited by

Michael E. Levinshtein

The Ioffe Institute, Russian Academy of Sciences

Sergey L. Rumyantsev

The Ioffe Institute, Russian Academy of Sciences

Michael S. Shur

Rensselaer Polytechnic Institute

A WILEY-INTERSCIENCE PUBLICATION

JOHN WILEY & SONS, INC.

New York / Chichester / Weinheim / Brisbane / Singapore / Toronto

Library of Congress Cataloging-in-Publication Data:

Properties of advanced semiconductor materials: GaN, AlN, InN, BN, SiC, SiGe / edited by M.E. Levinshtein, S.L. Rumyantsev, M.S. Shur.
    p. cm.
  Includes bibliographical references and index.
  ISBN 0-471-35827-4 (cloth : alk. paper)
  1. Semiconductors—Materials. I. Levinshtein, M.E. (Mikhail Efimovich) II. Rumyantsev, S.L. III. Shur, Michael.
  QC611.P78 2001
  537.6′226—dc21   00-042262

Printed in the United Kingdom.

10 9 8 7 6 5 4 3 2 1

To our parents

Contents

Contributors   **xiii**

Preface   **xv**

Chapter **1**   **Gallium Nitride (GaN)**   **1**
V. Bougrov, M. Levinshtein, S. Rumyantsev, and A. Zubrilov

  1.1.   Basic Parameters at 300 K / 1

  1.2.   Band Structure and Carrier Concentration / 3
      1.2.1.   Temperature Dependences / 4
      1.2.2.   Dependence on Hydrostatic Pressure / 6
      1.2.3.   Band Discontinuities at Heterointerfaces / 7
      1.2.4.   Effective Masses / 7
      1.2.5.   Donors and Acceptors / 8

  1.3.   Electrical Properties / 9
      1.3.1.   Mobility and Hall Effect / 9
      1.3.2.   Two-Dimensional Electron Gas Mobility at AlGaN/GaN Interface / 12
      1.3.3.   Transport Properties in High Electric Field / 13
      1.3.4.   Impact Ionization / 15
      1.3.5.   Recombination Parameters / 15

  1.4.   Optical Properties / 16

1.5.   Thermal Properties / 22

1.6.   Mechanical Properties, Elastic Constants, Lattice
       Vibrations, Other Properties / 24

References / 28

Chapter **2** **Aluminum Nitride (AlN)**   **31**
       Yu. Goldberg

2.1.   Basic Parameters at 300 K / 31

2.2.   Band Structure and Carrier Concentration / 33
       2.2.1.   Temperature Dependences / 33
       2.2.2.   Dependences on Hydrostatic Pressure / 34
       2.2.3.   Band Discontinuities at Heterointerfaces
                / 35
       2.2.4.   Effective Masses / 35
       2.2.5.   Donors and Acceptors / 36

2.3.   Electrical Properties / 37
       2.3.1.   Mobility and Hall Effect / 37
       2.3.2.   Recombination Parameters / 38

2.4.   Optical Properties / 39

2.5.   Thermal Properties / 41

2.6.   Mechanical Properties, Elastic Constants, Lattice
       Vibrations, Other Properties / 44

References / 46

Chapter **3** **Indium Nitride (InN)**   **49**
       A. Zubrilov

3.1.   Basic Parameters at 300 K / 49

3.2.   Band Structure and Carrier Concentration / 51
       3.2.1.   Temperature Dependences / 51
       3.2.2.   Dependence on Hydrostatic Pressure / 53
       3.2.3.   Band Discontinuities at Heterointerfaces
                / 53
       3.2.4.   Effective Masses / 53
       3.2.5.   Donors and Acceptors / 54

3.3.   Electrical Properties / 54
       3.3.1.   Mobility and Hall Effect / 54

3.3.2.   Transport Properties in High Electric
         Field / 56

3.3.3.   Impact Ionization / 57

3.3.4.   Recombination Parameters / 58

3.4.   Optical Properties / 58

3.5.   Thermal Properties / 61

3.6.   Mechanical Properties, Elastic Constants, Lattice
       Vibrations, Other Properties / 63

References / 65

Chapter 4   Boron Nitride (BN)   67
S. Rumyantsev, M. Levinshtein, A.D. Jackson, S.N. Mohammad,
G.L. Harris, M.G. Spencer, and M.S. Shur

4.1.   Basic Parameters at 300 K / 68

4.2.   Band Structure and Carrier Concentration / 70
       4.2.1.   Dependence on Hydrostatic Pressure / 72
       4.2.2.   Effective Masses / 73
       4.2.3.   Donors and Acceptors / 75

4.3.   Electrical Properties / 75

4.4.   Optical Properties / 76

4.5.   Thermal Properties / 80

4.6.   Mechanical Properties, Elastic Constants, Lattice
       Vibrations, Other Properties / 86

References / 91

Chapter 5   Silicon Carbide (SiC)   93
Yu. Goldberg, M. Levinshtein, and S. Rumyantsev

5.1.   Basic Parameters at 300 K / 93

5.2.   Band Structure and Carrier Concentration / 96
       5.2.1.   Temperature Dependences / 97
       5.2.2.   Dependence on Hydrostatic Pressure / 100
       5.2.3.   Energy Gap Narrowing at High Doping
               Levels / 101
       5.2.4.   Effective Masses / 102
       5.2.5.   Donors and Acceptors / 104

5.3. Electrical Properties / 106
    5.3.1. Mobility and Hall Effect / 106
    5.3.2. Transport Properties in High Electric
            Field / 111
    5.3.3. Impact Ionization / 114
    5.3.4. Recombination Parameters / 119

5.4. Optical Properties / 122

5.5. Thermal Properties / 132

5.6. Mechanical Properties, Elastic Constants, Lattice
      Vibrations, Other Properties / 138

References / 143

Chapter **6 Silicon-Germanium (Si$_{1-x}$Ge$_x$)**   **149**
      F. Schäffler

6.1. Basic Parameters in Unstrained Bulk Material at
      300 K / 151

6.2. Band Structure and Carrier Concentration / 154
    6.2.1. Temperature Dependences / 157
    6.2.2. Dependence of Energy Gap on Hydrostatic
            Pressure / 160
    6.2.3. Strain-Dependent Band Discontinuity / 160
    6.2.4. Effective Masses / 164

6.3. Electrical Properties / 168
    6.3.1. Mobility and Hall Effect / 168
    6.3.2. Two-Dimensional Electron Gas / 169
    6.3.3. Two-Dimensional Hole Gas / 171

6.4. Optical Properties / 173

6.5. Thermal Properties / 176

6.6. Mechanical Properties, Elastic Constants, Lattice
      Vibrations, Other Properties / 179

References / 186

Appendixes   **189**

1. Basic Physical Constants / 189

2. Periodic Table of the Elements / 190

3. Rectangular Coordinates for Hexagonal
   Crystal / 191

4. The First Brillouin Zone for Wurtzite Crystal / 191

5. Zinc Blende Structure / 192

6. The First Brillouin Zone for Zinc Blende
   Crystal / 192

Additional References   **193**

Contributors

V. Bourgrov, The Ioffe Institute, Russian Academy of Sciences, St. Petersburg, Russia

Yu. Goldberg, The Ioffe Institute, Russian Academy of Sciences, St. Petersburg, Russia

G.L. Harris, Howard University, Washington, D.C.

A.D. Jackson, Howard University, Washington, D.C.

M. Levinshtein, The Ioffe Institute, Russian Academy of Sciences, St. Petersburg, Russia

S.N. Mohammad, Howard University, Washington, D.C.

S. Rumyantsev, The Ioffe Institute, Russian Academy of Sciences, St. Petersburg, Russia

F. Schäffler, Johannes Kepler University, Linz, Austria

M. Shur, Rensselaer Polytechnic Institute, Troy, New York

M.G. Spencer, Howard University, Washington, D.C.

A. Zubrilov, The Ioffe Institute, Russian Academy of Sciences, St. Petersburg, Russia

Preface

This book contains information on emerging semiconductor materials systems ranging from wide band-gap semiconductors, such as SiC and AlN–GaN–InN–BN semiconductors, to SiGe compounds and heterostructures. At first glance, GaN, SiC, and SiGe, make strange bedfellows. However, what they have in common is an enormous amount of recent research on their properties and a shared need for a concise but fairly complete handbook describing their most important properties.

GaN-based and related materials have already demonstrated high-power and low-noise microwave performance that rivals and exceeds that of a much more mature GaAs technology. But the most exciting potential is probably in optoelectronic applications of GaN-based devices. A blue laser pioneered by Nakamura and efficient blue-, green-, amber-, and white-light-emitting diodes that emerged from pioneering work of Pankove hold promise of future solid-state replacements for inefficient and unreliable incandescent bulbs dating to Thomas Alva Edison time in terms of their basic design. Other potential applications of GaN-based technology include agriculture (for blue artificial plant lighting), medicine, biology, displays, radars, defense, and, possibly, many other areas still unforeseen. The first commercial SiC electronic devices have appeared on the market. High-voltage $4H$-SiC rectifier diodes with blocking voltages of up to 6.2 kV and thyristors with breakdown voltage of 3.1 kV have been successfully fabricated. SiGe

Heterojunction Bipolar Transistors find their niche in low-power, ultra-high-speed and wireless technologies.

It is our hope that our handbook will help researchers and engineers who will work hard in order to realize the enormous potential of these materials.

The technology and characterization of semiconductor materials discussed in this book is still emerging. This is reflected in the variety and complexity of experimental data for their material parameters. In many cases, the differences in the values of the material parameters measured by different authors exceed the stated accuracy of the measurements. In such cases, we chose the values that we believed to be more reliable.

We also tried to provide interpolation formulas in addition to graphs in order to make the information more quantitative. All in all, we tried to come up with a handbook that is useful for researchers or engineers working on these materials or related devices.

The discussion of semiconductor materials properties at a very basic level is given in *Introduction to Electronic Devices* by M.S. Shur (John Wiley & Sons, New York, 1996), which also contains convenient tables summarizing the basic semiconductor equations and the definitions of basic semiconductor parameters. A more detailed discussion is given in the two-volume set *Survey of Semiconductor Physics* by Karl Boer (Van Nostrand, New York, 1990).

In certain cases, tensor parameters (such as piezoelectric constants or elastic modulus) might be represented in different (but related) forms. R.E. Newnhams book *Structure–Property Relations* (Springer-Verlag, New York, 1975) can assist you in converting these parameters from the form given in this handbook to any other form.

Appendix 7 gives a list of additional key references dealing with the properties of the compounds discussed in this book. While we hope that this list will be helpful to the readers in need of more detailed information, it also makes the point that a concise handbook with carefully selected parameters might be very useful in order to find the needed information quickly. We are hoping that our book will serve this very purpose.

We are grateful to the contributors to this book, Drs. V. Bougrov, Yu. A. Goldberg, and A. Zubrilov (The Ioffe Institute, St. Petersburg, Russia), G.L. Harris, A.D. Jackson, S.N. Mohammad, and M.G. Spencer (Howard University, Washington, D.C.), and F. Schäffler (Johannes Kepler University, Linz, Austria), and to our colleagues at the A.F. Ioffe Institute and at Rensselaer Polytechnic Institute, Professors Leo Showalter, Glenn Slack, Drs. A. Andreev, A. Bykhovski, Pavel Ivanov, Sergey Karpov, W.V. Lundin, and Anatoly Strel'chuk, who helped us find additional information, made many excellent suggestions, and, in some cases, provided us with more accurate values of material parameters. If any credit is due us, they should share in this credit. The responsibility for possible errors is ours, and ours only.

Bertrand Russell once said, "A parameter in science is not a fact, but an instance." Therefore, we will greatly appreciate any comments or suggestions, which can be e-mailed to M.E. Levinshtein and S. Rumyantsev (melev@nimis.ioffe.rssi.ru) or to M.S. Shur (shurm@rpi.edu).

MICHAEL E. LEVINSHTEIN
SERGEY L. RUMYANTSEV
MICHAEL S. SHUR

St. Petersburg, Russia
Troy, New York
September 2000

PROPERTIES OF
ADVANCED
SEMICONDUCTOR
MATERIALS

Chapter 1

Gallium Nitride (GaN)

V. Bougrov, M. Levinshtein, S. Rumyantsev, and A. Zubrilov
The Ioffe Institute, St. Petersburg, Russia

1.1. BASIC PARAMETERS AT 300 K

	Wurtzite	Zinc Blende
Crystal structure	Wurtzite	Zinc Blende
Group of symmetry	$C_{6v}{}^4P6_3mc$	$T_d^2 - F\bar{4}3m$
Number of atoms in 1 cm^3		8.9×10^{22}
Debye temperature (K)		600
Density (g/cm^3)		6.15
Dielectric constant		
static	8.9	9.7
high frequency	5.35	5.3
Effective electron mass (in units of m_0)	0.20	0.13
Effective hole masses (in units of m_0)		
heavy	1.4	1.3
light	0.3	0.2
split-off band	0.6	0.3
Electron affinity (eV)		4.1

(Continued)

Properties of Advanced Semiconductor Materials, Edited by Levinshtein, Rumyantsev, Shur.
ISBN 0-471-35827-4 © 2001 John Wiley & Sons, Inc.

	Wurtzite	Zinc Blende
Lattice constants (Å)	$a = 3.189$	4.52
	$c = 5.186$	
Optical phonon energy (meV)	91.2	87.3
Band structure and carrier concentration		
Energy gap (eV)	3.39	3.2
Conduction band		
Energy separation between		
$\quad$ Γ valley and M–L valleys (eV)	$1.1 \div 1.9$	
$\quad$ M–L-valleys degeneracy	6	
Energy separation between		
$\quad$ Γ valley and A valleys (eV)	$1.3 \div 2.1$	
$\quad$ A-valley degeneracy	1	
Energy separation between		
$\quad$ Γ valley and X valleys (eV)		1.4
Energy separation between		
$\quad$ Γ valley and L valleys (eV)		$1.6 \div 1.9$
Effective conduction band density of states (cm^{-3})	2.3×10^{18}	1.2×10^{18}
Valence band		
Energy of spin–orbital splitting E_{so} (eV)	0.008	0.02
Energy of crystal-field splitting E_{cr} (eV)	0.04	
Effective valence band density of states (cm^{-3})	4.6×10^{19}	4.1×10^{19}
Electrical properties		
Breakdown field $(\mathrm{V\ cm}^{-1})$	$\sim 5 \times 10^6$	$\sim 5 \times 10^6$
Mobility $(\mathrm{cm}^2\ \mathrm{V}^{-1}\ \mathrm{s}^{-1})$		
$\quad$ electrons	≤ 1000	≤ 1000
$\quad$ holes	≤ 200	≤ 350
Diffusion coefficient $(\mathrm{cm}^2\ \mathrm{s}^{-1})$		
$\quad$ electrons	25	25
$\quad$ holes	5	9
Electron thermal velocity $(\mathrm{m\ s}^{-1})$	2.6×10^5	3.2×10^5
Hole thermal velocity $(\mathrm{m\ s}^{-1})$	9.4×10^4	9.5×10^4

(Continued)

	Wurtzite	Zinc Blende
Optical properties		
Infrared refractive index		2.3
Radiative recombination coefficient (cm^3 s^{-1})		10^{-8}
Thermal and mechanical properties		
Bulk modulus (dyn cm^{-2})		20.4×10^{11}
Melting point (°C)		2500 (see Fig. 1.5.5)
Specific heat (J g^{-1} $°C^{-1}$)		0.49
Thermal conductivity (W cm^{-1} $°C^{-1}$)		1.3
Thermal diffusivity (cm^2 s^{-1})		0.43
Thermal expansion, linear ($°C^{-1}$)	$\alpha_a = 5.59 \times 10^{-6}$ $\alpha_c = 3.17 \times 10^{-6}$	

1.2. BAND STRUCTURE AND CARRIER CONCENTRATION

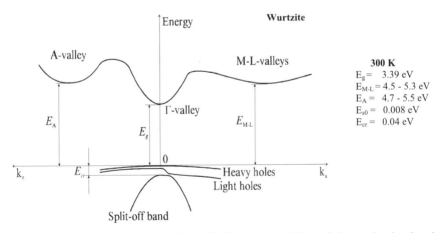

300 K
$E_g =$ 3.39 eV
$E_{M\text{-}L} =$ 4.5 - 5.3 eV
$E_A =$ 4.7 - 5.5 eV
$E_{s0} =$ 0.008 eV
$E_{cr} =$ 0.04 eV

FIG. 1.2.1. Band structure of wurtzite GaN. Important minima of the conduction band and maxima of the valence band. Valence band of wurtzite GaN has three splitted bands. This splitting results from spin–orbit interaction and from crystal symmetry. [For details see Suzuki and Uenoyama (1995)].

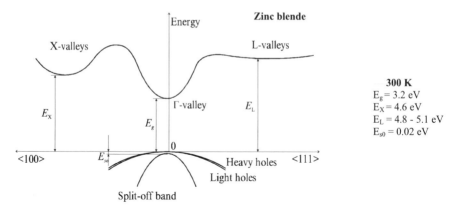

300 K
$E_g = 3.2$ eV
$E_X = 4.6$ eV
$E_L = 4.8 - 5.1$ eV
$E_{s0} = 0.02$ eV

FIG. 1.2.2. Band structure of zinc blende GaN. Important minima of the conduction band and maxima of the valence band.

1.2.1. Temperature Dependences

Temperature dependence of energy gap:

$$E_g = E_g(0) - 7.7 \times 10^{-4} \times \frac{T^2}{T + 600} \ (\text{eV}) \qquad (1.2.1)$$

where T is temperature in degrees K.

$$E_g(0) \ (\text{eV}): \qquad 3.47 \ (\text{wurtzite}) \qquad 3.28 \ (\text{zinc blende})$$

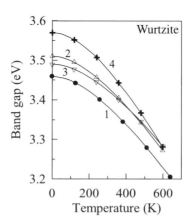

FIG. 1.2.3. The temperature dependences of wurtzite GaN band gap. GaN samples were grown on different substrates using different techniques. Experimental data are taken from four different works.

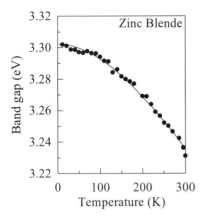

FIG. 1.2.4. The temperature dependence of cubic GaN band gap. GaN films were grown on MgO (1×1) substrates [Ramirez-Flores et al. (1994)].

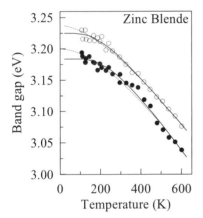

FIG. 1.2.5. The temperature dependences of cubic GaN band gap. GaN films were grown on Si (100) substrates. The dependences were extracted from pseudodielectric-function spectrum using two different theoretical models [Petalas et al. (1995)].

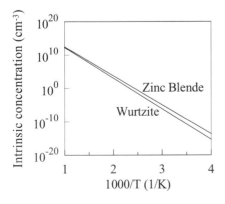

FIG. 1.2.6. The temperature dependences of the intrinsic carrier concentration.

Intrinsic carrier concentration:

$$n_i = (N_c \cdot N_v)^{1/2} \exp\left(-\frac{E_g}{2k_B T}\right) \qquad (1.2.2)$$

Effective density of states in the conduction band: N_c

Wurtzite

$$N_c \cong 4.82 \times 10^{15} \cdot \left(\frac{m_\Gamma}{m_0}\right)^{3/2} \cdot T^{3/2} \; (\text{cm}^{-3})$$

$$\cong 4.3 \times 10^{14} \times T^{3/2} \; (\text{cm}^{-3}) \qquad (1.2.3)$$

Zinc Blende

$$N_c \cong 4.82 \times 10^{15} \cdot \left(\frac{m_\Gamma}{m_0}\right)^{3/2} \cdot T^{3/2} \; (\text{cm}^{-3})$$

$$\cong 2.3 \times 10^{14} \times T^{3/2} \; (\text{cm}^{-3}) \qquad (1.2.4)$$

Effective density of states in the valence band: N_v

Wurtzite

$$N_v = 8.9 \times 10^{15} \times T^{3/2} \; (\text{cm}^{-3}) \qquad (1.2.5)$$

Zinc Blende

$$N_v = 8.0 \times 10^{15} \times T^{3/2} \; (\text{cm}^{-3}) \qquad (1.2.6)$$

1.2.2. Dependence on Hydrostatic Pressure

For wurtzite GaN:

$$E_g = E_g(0) + 4.2 \times 10^{-3} P - 1.8 \times 10^{-5} P^2 \; (\text{eV}) \qquad (1.2.7)$$

where P is pressure in kbar [Morkoc et al. (1994), Akasaki and Amano (1994a)].

1.2.3. Band Discontinuities at Heterointerfaces

AlN/GaN (0001) (wurtzite) [Martin et al. (1996)]
 Conduction band discontinuity $\Delta E_c = 2.0$ eV
 Valence band discontinuity $\Delta E_v = 0.7$ eV

InN/GaN (wurtzite) [Martin et al. (1996)]
 Conduction band discontinuity $\Delta E_c = 0.43$ eV
 Valence band discontinuity $\Delta E_v = 1.0$ eV

GaAs/GaN (cubic) [Ding et al. (1997)]
 Valence band discontinuity $\Delta E_v = 1.84$ eV

1.2.4. Effective Masses

Electrons

For wurtzite crystal structure the surfaces of equal energy in Γ valley should be ellipsoids, but effective masses in z direction and perpendicular directions are estimated to be approximately the same:

 For wurtzite GaN: $m_\Gamma = 0.20 m_0$.
 For zinc blende GaN: $m_\Gamma = 0.13 m_0$.

Holes

Experimental data for wurtzite crystals give the value of hole effective mass about $1.0 m_0$. This value is less than the calculated one. Here we give the calculated values. For zinc blende crystals, only calculated data are available [Leszczynski et al. (1996), Fan et al. (1996)].

	Wurtzite	Zinc Blende
Effective mass of density of state m_v	$1.5 m_0$	$1.4 m_0$
heavy holes	$m_{hh} = 1.4 m_0$	$m_{hh} = 1.3 m_0$
	$m_{hhz} = 1.1 m_0$	$m_{[100]} = 0.8 m_0$
	$m_{hh\perp} = 1.6 m_0$	$m_{[111]} = 1.7 m_0$
light holes	$m_{lh} = 0.3 m_0$	$m_{lh} = 0.19 m_0$
	$m_{lhz} = 1.1 m_0$	$m_{[100]} = 0.21 m_0$
	$m_{lh\perp} = 0.15 m_0$	$m_{[111]} = 0.18 m_0$
split-off holes	$m_{sh} = 0.6 m_0$	$m_{sh} = 0.33 m_0$
	$m_{shz} = 0.15 m_0$	$m_{[100]} = 0.33 m_0$
	$m_{sh\perp} = 1.1 m_0$	$m_{[111]} = 0.33 m_0$

1.2.5. Donors and Acceptors

Wurtzite

Ionization Energies of Shallow Donors (eV)

Si	0.012–0.02
Native defect level (V_N)	0.03

Ionization Energies of Shallow Acceptors (eV)
[Strite and Morkoc (1992), Akasaki and Amano (1994b)]

Mg	0.14–0.21
Zn	0.21
Native defect level (V_{Ga})	0.14

The Most Important Levels for Wurtzite GaN

Impurity or Defect	Ga Site	N Site
Donors, Ionization Energy $(E_t - E_c)$ (eV)		
Si	0.012–0.02	
V_N (vacancy)		0.03; 0.1
C	0.11–0.14	
Mg		0.26; 0.6
Acceptors, Ionization Energy $(E_v - E_t)$ (eV)		
V_{Ga} (vacancy)	0.14	
Mg	0.14–0.21	
Si		0.19
Zn	0.21–0.34	
Hg	0.41	
Cd	0.55	
Be	0.7	
Li	0.75	
C		0.89
Ga		0.59–1.09

Zinc Blende

For cubic GaN, only calculated data are available [Neugebauer and Van de Walle (1994), Boguslawski et al. (1995), Mattila et al. (1996), Boguslawski and Bernholc (1996), Gorczyca et al. (1997)].

1.3. ELECTRICAL PROPERTIES

1.3.1. Mobility and Hall Effect

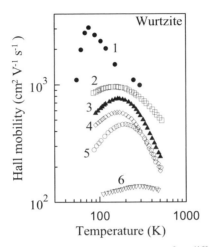

FIG. 1.3.1. Electron Hall mobility versus temperature for different doping levels and different degrees of compensation $\theta = N_a/N_d$ (wurtzite GaN). 1—unintentionally doped [Nakamura et al. (1992)]; 2—$N_d = 3.1 \times 10^{17}$ cm^{-3}, $\theta < 0.03$; 3—$N_d = 1.1 \times 10^{17}$ cm^{-3}, $\theta \approx 0.3$; 4—$N_d = 2.3 \times 10^{17}$ cm^{-3}, $\theta < 0.04$; 5—$N_d = 7.4 \times 10^{17}$ cm^{-3}, $\theta < 0.01$; 6—The concentration of introduced Si $N_{do} = 2 \times 10^{19}$ cm^{-3}. Curves 2–6 after Gotz et al. (1996). The calculations of electron mobility as a function of temperature for different doping levels and compensation ratios can be found in Chin et al. (1994).

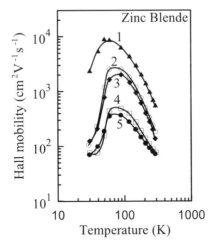

FIG. 1.3.2. Electron Hall mobility versus temperature for different doping levels (cubic GaN). Electron concentration at room temperature: 1: 1.5×10^{18} cm^{-3}; 2: 1.3×10^{19} cm^{-3}; 3: 2.8×10^{19} cm^{-3}; 4: 1.5×10^{20} cm^{-3}, 5: 3×10^{20} cm^{-3} [Kim et al. (1994)].

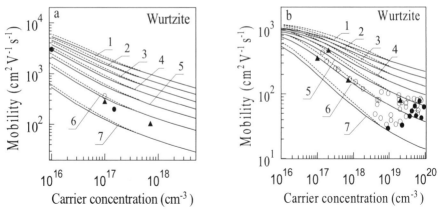

FIG. 1.3.3. The electron drift (solid curves) and Hall (dashed curves) mobility of wurtzite GaN calculated as a function of carrier concentration at different compensation ratios θ : 1—$\theta = 0$; 2—$\theta = 0.15$; 3—$\theta = 0.30$; 4—$\theta = 0.45$; 5—$\theta = 0.60$; 6—$\theta = 0.75$; 7—$\theta = 0.90$. (a) $T = 77$ K, (b) $T = 300$ K. Experimental data are taken from four different papers [Chin et al. (1994)].

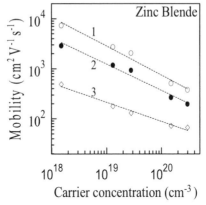

FIG. 1.3.4. The electron Hall mobility of cubic GaN at three temperatures as a function of carrier concentration. T (K): 1—80; 2—150; 3—300. The data are taken from Kim et al. (1994).

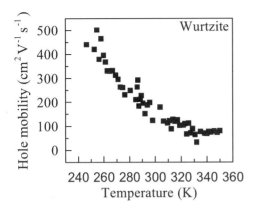

FIG. 1.3.5. Hole Hall mobility versus temperature for wurtzite GaN. Hole concentration at $T = 300$ K $p \approx 4 \times 10^{12}$ cm^{-3} [Rubin et al. (1994)].

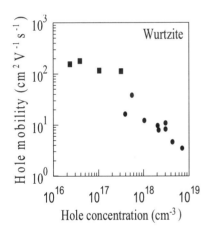

FIG. 1.3.6. Hole Hall mobility versus hole concentration for wurtzite GaN at $T = 300$ K [Gaskill et al. (1995)].

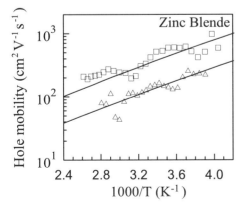

FIG. 1.3.7. Temperature dependence of the hole Hall mobility for two samples. Cubic GaN with hole concentration $p = (0.6 - 1) \times 10^{13}$ cm^{-3} at $T = 300$ K [As et al. (1996)].

1.3.2. Two-Dimensional Electron Gas Mobility at AlGaN/GaN Interface

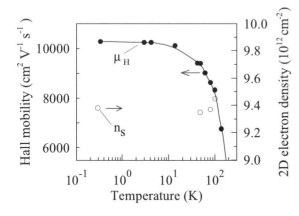

FIG. 1.3.8. Electron Hall mobility and sheet concentration as a function of temperature for two-dimensional gas in AlGaN/GaN heterostructure grown on 6H–SiC substrate [Gaska et al. (1998)].

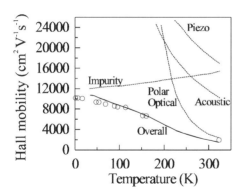

FIG. 1.3.9. Measured (open circles) and calculated (solid lines) Hall mobility as a function of temperature for the same sample as in Fig. 1.3.8. Electron scattering processes by optical and acoustic phonons, piezoelectric, and impurity scattering were taken into account [Gaska et al. (1998)].

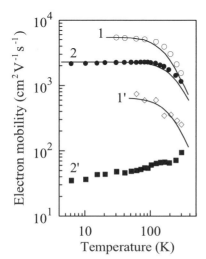

FIG. 1.3.10. Electron Hall mobility as a function of temperature for two AlGaN/GaN heterostructures $(1, 2)$ and related GaN layers $(1', 2')$ grown on sapphire. Electron concentrations in the two-dimensional electron gas and GaN layers at room temperature are: 1: $n_{2DEG} \approx 7 \times 10^{12}$ cm^{-2}, 2: $n_{2DEG} \approx 7.5 \times 10^{12}$ cm^{-2}, 1': $n \approx 7 \times 10^{16}$ cm^{-3}, 2': $n \approx 1.5 \times 10^{18}$ cm^{-3} [Dziuba et al. (1997)].

1.3.3. Transport Properties in High Electric Field

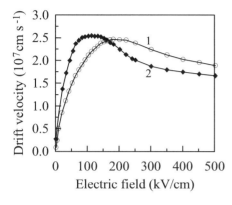

FIG. 1.3.11. Calculated steady-state drift velocity in wutzite (curve 1) and zinc blende (curve 2) GaN as function of electric field applied along (100) direction in the zinc blende and along (1010) direction in the wurtzite GaN [Kolnik et al. (1995); see also Albrecht et al. (1998)].

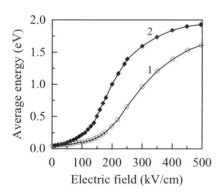

FIG. 1.3.12. Calculated average electron energy as a function of electric field in wurtzite (curve 1) and zinc blende (curve 2) GaN [Kolnik et al. (1995)].

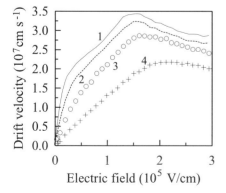

FIG. 1.3.13. Calculated field dependences of electron drift velocity in wurtzite GaN for different temperatures. $N_d = n = 10^{17}$ cm^{-3}. T (K): 1—77, 2—300, 3—500, 4—1000 [Bhapkar and Shur (1997)].

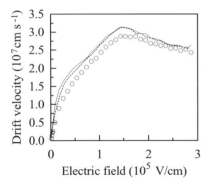

FIG. 1.3.14. Calculated field dependences of electron drift velocity in wurtzite GaN at $T = 300$ K for different doping levels: solid line, $N_d = n = 10^{16}$ cm^{-3}, dashed line, $N_d = n = 10^{17}$ cm^{-3}; circles $N_d = n = 10^{18}$ cm^{-3} [Bhapkar and Shur (1997)].

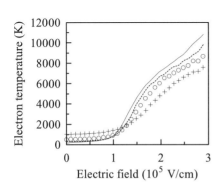

FIG. 1.3.15. Electron temperature as function of electric field in wurtzite GaN, $N_d = n = 10^{17}$ cm^{-3}. Solid line, $T = 77$ K; dashed line, $T = 300$ K; circles, $T = 500$ K; pluses, $T = 1000$ K [Bhapkar and Shur (1997)].

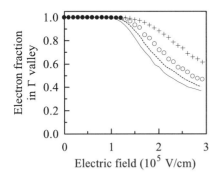

FIG. 1.3.16. Electron fraction in central valley of wurtzite GaN as a function of electric field. Solid line, $T = 77$ K; dashed line, $T = 300$ K; circles, $T = 500$ K; pluses, $T = 1000$ K [Bhapkar and Shur (1997)].

1.3.4. Impact Ionization

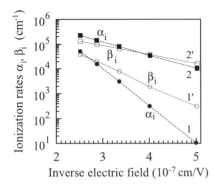

FIG. **1.3.17.** Calculated impact ionization rates as a function of inverse electric field for electrons (α_i) and holes (β_i) in wurtzite and zinc blende GaN. 300 K. 1,1′—wurtzite GaN; 2, 2′—zinc blende GaN [Oguzman et al. (1997)].

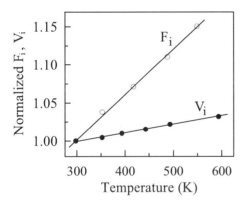

FIG. **1.3.18.** Normalized breakdown electric field (F_i) and relative breakdown voltage (V_i) as a function of temperature measured in wurtzite GaN p^+–p–n diodes. $F_i(300 \text{ K}) \approx (1\text{–}2) \times 10^6$ V/cm. $V_i(300 \text{ K}) \approx 42$ V [Osinsky et al. (1998)].

1.3.5. Recombination Parameters

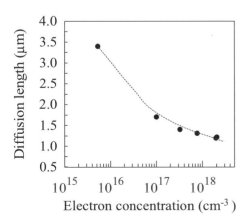

FIG. **1.3.19.** The dependence of hole diffusion length versus electron concentration for wurtzite GaN [Chernyak et al. (1996)].

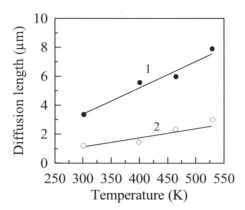

FIG. 1.3.20. The temperature dependences of hole diffusion lengths for two values of electron concentration for wurtzite GaN at 300 K. 1—$n = 5 \times 10^{15}$ cm^{-3}, 2— $n = 2 \times 10^{18}$ cm^{-3} [Chernyak et al. (1996)].

Radiative recombination coefficient at 300 K [Muth et al. (1997)]: 1.1×10^{-8} cm^3 s^{-1}.

1.4. OPTICAL PROPERTIES

Infrared refractive index (300 K) (wurtzite and zinc blende GaN): $n_\infty \approx 2.3$

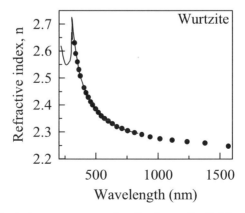

FIG. 1.4.1. Refractive index n versus wavelength of wurtzite GaN on sapphire at 300 K [Yu et al. (1997)].

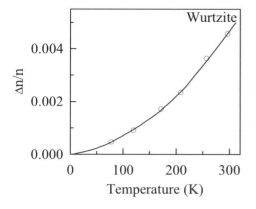

FIG. 1.4.2. Relative variation of the long-wavelength refractive index of wurtzite GaN with temperature [Ejder (1971)].

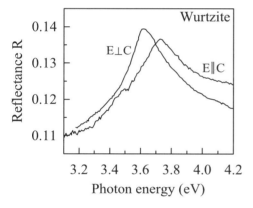

FIG. 1.4.3. Reflectance R as a function of photon energy for GaN single crystals (platelets). $T = 2$ K [Dingle et al. (1971)].

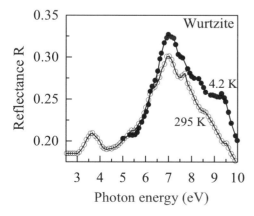

FIG. 1.4.4. Reflectance R as a function of photon energy for two temperatures [Bloom et al. (1974)].

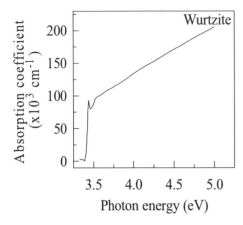

FIG. 1.4.5. The absorption coefficient versus photon energy for GaN layer grown on sapphire. $T = 293$ K [Muth et al. (1997)].

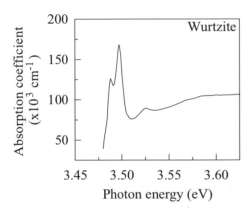

FIG. 1.4.6. The absorption coefficient versus photon energy for GaN layer grown on sapphire. $T = 77$ K [Muth et al. (1997)].

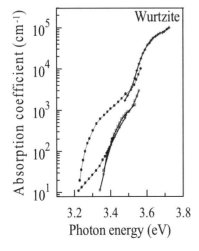

FIG. 1.4.7. Axial absorption spectra for several wurtzite GaN single crystals (platelets) at low temperature (∼5 K), E⊥C [Dingle et al. (1971)].

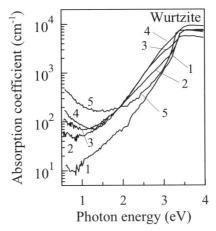

FIG. 1.4.8. The absorption coefficient versus photon energy for wurtzite GaN deposited on r-plane sapphire at different electron concentrations. $T = 300$ K. n_0 (cm^{-3}): 1– 2×10^{16}, 2–2.8×10^{17}, 3–5×10^{17}, 4–2.3×10^{18}, 5–2×10^{19} [Ambacher et al. (1996)].

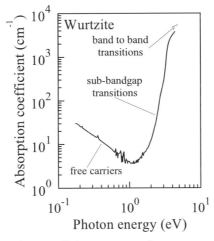

FIG. 1.4.9. The absorption coefficient versus photon energy for wurtzite GaN (MOCVD on c-plane sapphire). $T = 300$ K, $n_0 \approx 10^{17}$ cm^{-3} [Ambacher et al. (1996)].

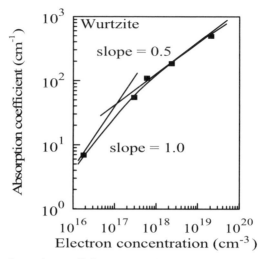

FIG. 1.4.10. The absorption coefficient versus electron concentration at photon energy $E_{ph} = 0.6$ eV (free carrier absorption) $T = 300$ K [Ambacher et al. (1996)].

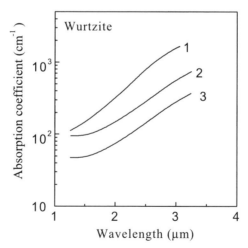

FIG. 1.4.11. The absorption coefficient versus wavelength for wurtzite GaN. Electron concentration n_0 (cm^{-3}): 1–6.3 × 10^{19}, 2–2.9 × 10^{19}, 3–1.8 × 10^{19} [Cunningham et al. (1972)].

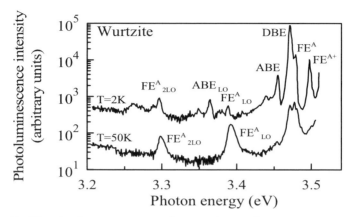

FIG. 1.4.12. Photoluminescence spectra for $T = 2$ K and $T = 50$ K. Wurtzite GaN bulk-like sample with the thickness of 500 µm [Monemar et al. (1996)].

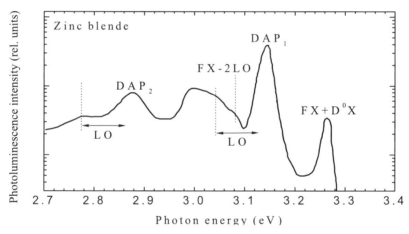

FIG. 1.4.13. Photoluminescence spectrum for cubic GaN grown on GaAs. $T = 1.8$ K [Holst et al. (1998)].

1.5. THERMAL PROPERTIES

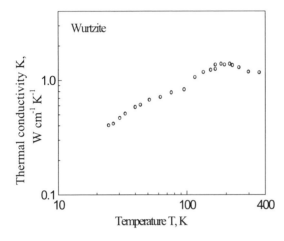

FIG. 1.5.1. Temperature dependence of thermal conductivity for wurtzite GaN [Sichel and Pankove (1977)] Reprinted from *J. Phys. Chem. Solids* **38**, Sichel, E.K., and Pankove, J.I., "Thermal conductivity of GaN. 25–360 K," p. 330, copyright © 1977, with permission from Elsevier Science.

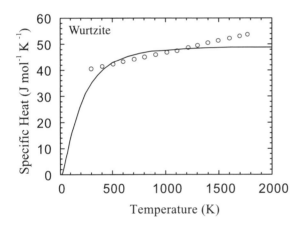

FIG. 1.5.2. The calculated (solid line) and measured (open circles) specific heat for GaN [Nipko et al. (1998)]. Experimental points are taken from Barin et al. (1977).

The specific heat C_p of wurtzite GaN at constant pressure for 298 K $< T <$ 1773 K [Barin et al. (1977)]:

$$C_p = 38.1 + 8.96 \times 10^{-3}T \qquad (\text{J mol}^{-1}\,\text{K}^{-1})$$

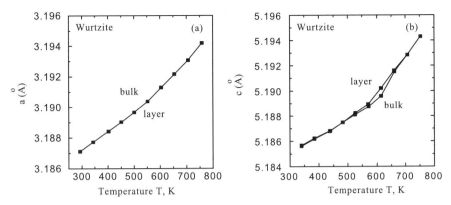

FIG. 1.5.3. Temperature dependences of wurtzite GaN lattice constants **a** and **c** [Leszc-zynski et al. (1994)].

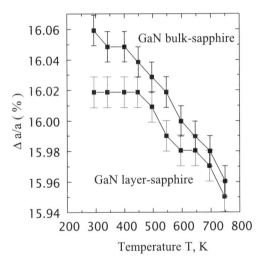

FIG. 1.5.4. The temperature dependences of the parallel lattice mismatch between GaN layer and sapphire substrate [Leszczynski et al. (1994)].

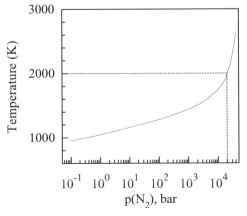

FIG. 1.5.5. Equilibrium N_2 pressure over GaN. Dashed lines indicate the maximum pressure and temperarure available in the experimental system [Porowski (1997)].

1.6. MECHANICAL PROPERTIES, ELASTIC CONSTANTS, LATTICE VIBRATIONS, OTHER PROPERTIES

Density: 6.15 g cm^{-3}

Surface microhardness
(using Knoop's pyramid test): 1200–1700 kg mm^{-2}
[Nikolaev et al. (1998), see also
Drory et al. (1996)]

Wurtzite

Elastic constants at 300 K [Polian et al. (1996)]:

C_{11}	390 ± 15 GPa
C_{12}	145 ± 20 GPa
C_{13}	106 ± 20 GPa
C_{33}	398 ± 20 GPa
C_{44}	105 ± 10 GPa

[see also Wright (1997)]

Bulk modulus B_s (compressibility^{-1}):

$$B_s = \frac{C_{33}(C_{11} + C_{12}) - 2(C_{13})^2}{C_{11} + C_{12} - 4C_{13} + 2C_{33}}, \qquad B_s = 210 \pm 10 \text{ GPa}$$

Acoustic Wave Speeds

Wave Propagation Direction	Wave Character	Expression for Wave Speed	Wave Speed (in units of 10^5 cm/s)
[001]	V_L (longitudinal)	$(C_{33}/\rho)^{1/2}$	8.04
	V_T (transverse)	$(C_{44}/\rho)^{1/2}$	4.13
[100]	V_L (longitudinal)	$(C_{11}/\rho)^{1/2}$	7.96
	V_T (transverse, polarization along [001])	$(C_{44}/\rho)^{1/2}$	4.13
	V_T (transverse, polarization along [010])	$[(C_{11} - C_{12})/2\rho]^{1/2}$	6.31

For definitions of the crystallographic directions see Appendix 3. For other details see R. Truell et al. (1969).

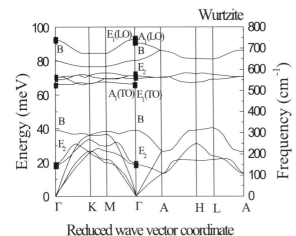

FIG. 1.6.1. Calculated dispersion curves for acoustic and optical branch phonons for wurtzite GaN [Siegle et al. (1997)].

Frequencies (in cm^{-1}) and Symmetries of the Strongest Modes Found in the Second-Order Raman Spectra of Hexagonal GaN and Their Assignments [Siegle et al. (1997)].

Frequency (cm^{-1})		Symmetry	Process	Point in the BZ
	317	A_1	Acoustic, overtone	H
	410	A_1	Acoustic, overtone	A, K
	420	A_1, E_2	Acoustic, overtone	M
	533	A_1	First-order process	$\Gamma = A_1(\mathrm{TO})$
	560	E_1	First-order process	$\Gamma = E_1(\mathrm{TO})$
	569	E_2	First-order process	$\Gamma = E_2(\mathrm{high})$
	640	A_1	Overtone	$\Gamma = [B]^2, L$
	735	A_1	First-order process	$\Gamma = A_1(\mathrm{LO})$
	742	E_1	First-order process	$\Gamma = E_1(\mathrm{LO})$
	855	A_1, E_1, E_2	Acoustic–optical combination	
	915	A_1	Acoustic–optical combination	
	1000	$A_1, (E_2)$	Acoustic–optical combination	
	1150	A_1	Optical overtone	
	1280	$A_1, (E_1)$	Optical combination	
	1289	E_2	Optical combination	
	1313	$A_1, (E_1, E_2)$	Optical combination	
	1385	A_1	Optical overtone	A, K
	1465	A_1	Optical overtone	$\Gamma = A_1(\mathrm{LO})^2$
Cutoff	1495	A_1, E_2	Optical overtone	$\Gamma = E_1(\mathrm{LO})^2$

Main Phonon Frequencies (in units of cm^{-1})
[Siegle et al. (1997), Akasaki and Amano (1994),
Karch et al. (1998), and Zi et al. (1996)]

$A_1 - LO$	710–735
$A_1 - TO$	533–534
$E_1 - LO$	741–742
$E_1 - TO$	556–559
E_2 (low)	143–146
E_2 (high)	560–579

Piezoelectric Constants [Bykhovski et al. (1997)]

e_{15}	-0.30 C m^{-2}
e_{31}	-0.33 C m^{-2}
e_{33}	0.65 C m^{-2}

Zinc Blende

Elastic constants at 300 K [Wright (1997)]:

C_{11}	293 GPa
C_{12}	159 GPa
C_{44}	155 GPa

Bulk modulus B_s (compressibility^{-1}):

$$B_s = \frac{C_{11} + 2C_{12}}{3}, \qquad B_s = 204 \text{ GPa}$$

Anisotropy factor:

$$A = \frac{C_{11} - C_{12}}{2C_{44}}, \qquad A = 0.43$$

Shear modulus:

$$C' = (C_{11} - C_{12})/2, \qquad C' = 67 \text{ GPa}$$

[100] Young's modulus Y_0:

$$Y_0 = \frac{(C_{11} + 2C_{12})(C_{11} - C_{12})}{(C_{11} + C_{12})}, \qquad Y_0 = 181 \text{ GPa}$$

[100] Poisson ratio σ_0:

$$\sigma_0 = \frac{C_{12}}{C_{11} + C_{12}}, \qquad \sigma_0 = 0.352$$

Acoustic Wave Speeds

Wave Propagation Direction	Wave Character	Expression for Wave Speed	Wave Speed (in units of 10^5 cm/s)
[100]	V_L	$(C_{11}/\rho)^{1/2}$	6.9
	V_T	$(C_{44}/\rho)^{1/2}$	5.02
[110]	V_1	$[(C_{11} + C_{12} + 2C_{44})/2\rho)]^{1/2}$	7.87
	$V_{t\|}$	$V_{t\|} = V_T = (C_{44}/\rho)^{1/2}$	5.02
	$V_{t\perp}$	$[(C_{11} - C_{12})/2\rho)]^{1/2}$	3.3
[111]	V_1'	$[(C_{11} + 2C_{12} + 4C_{44})/3\rho)]^{1/2}$	8.17
	V_t'	$[(C_{11} - C_{12} + C_{44})/3\rho)]^{1/2}$	3.96

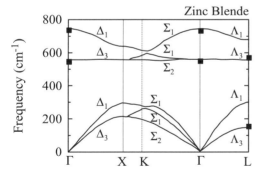

FIG. 1.6.2. Calculated dispersion curves for acoustic and optical branch phonons for cubic GaN [Zi et al. (1996)].

Phonon Frequencies (cm^{-1}) [Zi et al. (1996)]

LO (Γ)	748	TA (X)	207
TO (Γ)	562	LO (L)	675
LO (X)	639	TO (L)	554
TO (X)	558	LA (L)	296
LA (X)	286	TA (L)	144

Piezoelectric constant e_{14}:   0.4 C m^{-2} [Shur et al. (1996)]

REFERENCES

Akasaki, I., and H. Amano, in *Properties of Group III Nitrides* (edited by Edgar, J.H.), EMIS Datareviews Series, No. 11, (1994a), an INSPEC publication, pp. 30–34.

Akasaki, I., and H. Amano, *J. Electrochem. Soc.* **141**, 8 (1994b), 2266–2271.

Albrecht, J.D., R.P. Wang, P.P. Ruden, M. Farahmand, and K.F. Brennan, *J. Appl. Phys.* **83**, 9 (1998), 4777–4781.

Ambacher, O., W. Rieger, P. Ansmann, H. Angerer, T.D. Moustakas, and M. Stutzman, *Sol. State Commun.* **97**, 5 (1996), 365–370.

As, D.J., D. Schikora, A. Greiner, M. Lubbers, J. Mimkes, and K. Lischka, *Phys. Rev. B* **54**, 16 (1996), R11118–R11121.

Barin, I., O. Knacke, and O. Kubaschewski, *Thermochemical Properties of Inorganic Substances*, Springer-Verlag, Berlin, 1977.

Bhapkar, U.V., and M.S. Shur, *J. Appl. Phys.* **82**, 4 (1997), 1649–1655.

Bloom, S., G. Harbeke, E. Meier, and I.B. Ortenburger, *Phys. Stat. Solidi* **66** (1974), 161–168.

Boguslawski, P., E.L. Briggs, and J. Bernholc, *Phys. Rev. B* **51**, 23 (1995), 17255–17258.

Boguslawski, P., and J. Bernholc, *Acta Phys. Pol A* **90** (1996), 735.

Bykhovski, A.D., B.L. Gelmont, and M.S. Shur, *J. Appl. Phys.* **81**, 9 (1997), 6332–6338.

Chernyak, L., A. Osinsky, H. Temkin, J.W. Yang, Q. Chen, and M.A. Khan, *Appl. Phys. Lett.* **69**, 17 (1996), 2531–2533.

Chin, V.W.L., T.L. Tansley, and T. Osotchan, *J. Appl. Phys.* **75**, 11 (1994), 7365–7372.

Cunningham, R.D., R.W. Brander, N.D. Knee, and D.K. Wickenden, *J Lumin.* **5** (1972), 21–31.

Ding, S.A., S.R. Barman, K. Horn, H. Yang, B.B. Yang, O. Brandt, and K. Ploog, *Appl. Phys. Lett.* **70**, 18 (1997), 2407–2409.

Dingle, R., D.D. Sell, S.E. Stokowski, P.J. Dean, and R.B. Zetterstrom, *Phys. Rev. B* **3**, 2 (1971), 497–500.

Drory, M.D., J.W. Ager, T. Suski, I. Grzegory, and S. Porowski, *Appl. Phys. Lett.* **69**, 26 (1996), 4044–4046.

Dziuba, Z., J. Antoszewski, J.M. Dell, L. Faraone, P. Kozodoy, S. Keller, B. Keller, S.P. DenBaars, and U.K. Mishra, *J. Appl. Phys.* **82**, 6 (1997), 2996–3002.

Ejder, E., *Phys. Stat. Solidi* (a) **6** (1971), 445–448.

Fan, J.W., M.F. Li, T.C. Chong, and J.B. Xia, *J. Appl. Phys.* **79**, 1 (1996), 188–194.

Gaska, R., J.W. Yang, A. Osinsky, Q. Chen, M.A. Khan, A.O. Orlov, G.L. Snider, and M.S. Shur, *Appl. Phys. Lett.* **72**, 6 (1998), 707–709.

Gaskill, D.K., L.B. Rowland, and K. Doverspike, Electrical transport properties of AlN, GaN and AlGaN, in *Properties of Group III Nitrides* (edited by Edgar, J.), EMIS Datareviews Series, No. 11, 1995, pp. 101–116.

Gorczyca, I., A. Svane, and N.E. Christensen, *Internet J. Nitride Semicond. Res.* **2**, Article 18 (1997).

Gotz, W., N.M. Johnson, C. Chen, H. Liu, C. Kuo, and W. Imler, *Appl. Phys. Lett.* **68**, 22 (1996), 3144–3146.

Holst, J., L. Eckey, A. Hoffmann, I. Broser, B. Schottker, D.J. As, D. Schikora, and K. Lischka, *Appl. Phys. Lett.* **72**, 12 (1998), 1439–1441.

Karch, K., J.M. Wagner, and F. Bechstedt, *Phys. Rev. B* **57**, 12 (1998), 7043–7049.

Kim, J.G., A.C. Frenkel, H. Liu, and R.M. Park, *Appl. Phys. Lett.* **65** (1994), 91–93.

Kolnik, J., I.H. Oguzman, K.F. Brennan, R. Wang, P.P. Ruden, and Y. Wang, *J. Appl. Phys.* **78**, 2 (1995), 1033–1038.

Leszczynski, M., T. Suski, H. Teisseyre, P. Perlin, I. Grzegory, J. Jun, S. Porowski, and T.D. Moustakas, *J. Appl. Phys.* **76**, 8 (1994), 4909–4911.

Leszczynski, M., H. Teisseyre, T. Suski, I. Grzegory, M. Bockowski, J. Jun, S. Porowski, K. Pakula, J.M. Baranowski, C.T. Foxon, and T.S. Cheng, *Appl. Phys. Lett.* **69**, 1 (1996), 73–75.

Martin, G., A. Botchkarev, A. Rockett, and H. Morkoc, *Appl. Phys. Lett.* **68**, 18 (1996), 2541–2543.

Mattila, T., A.P. Seitsonen, and R.M. Nieminen, *Phys. Rev. B* **54**, 3 (1996), 1474–1477.

Monemar, B., J.P. Bergman, H. Amano, I. Akasaki, T. Detchprohm, K. Hiramatsu, and N. Sawaki, International Symposium on Blue Laser and Light Emiting Diodes, Chiba University, Japan, March 5–7, 1996.

Morkoc, H., S. Strite, G.B. Gao, M.E. Lin, B. Sverdlov, and M. Burns, *J. Appl. Phys.* **76**, 3 (1994), 1363–1397.

Muth, J.F., J.H. Lee, I.K. Shmagin, R.M. Kolbas, H.C. Casey, Jr., B.P. Keller, U.K. Mishra, and S.P. DenBaars, *Appl. Phys. Lett.* **71**, 18 (1997), 2572–2574.

Nakamura, S., T. Mukai, and M. Senoh, *Jpn. J. Appl. Phys.* **31** (1992), 2883–2888.

Neugebauer, J., and C.G. Van de Walle, *Phys. Rev. B* **50**, 11 (1994), 8067–8070.

Nikolaev, V., V. Shpeizman, and B. Smirnov, The Second Russian Workshop "GaN, InN, and AlN-structures and devices." St. Petersburg, St. Petersburg Technical University Russia, June 2, 1998.

Nipko, J.C., C.-K. Loong, C.M. Balkas, and R.F. Davis, *Appl. Phys. Lett.* **73**, 1 (1998), 34–35.

Oguzman, I.H., E. Bellotti, K.F. Brennan, J. Kolnik, R. Wang, and P.P. Ruden, *J. Appl. Phys.* **81**, 12 (1997), 7827–7834.

Osinsky, A., M.S. Shur, R. Gaska, and Q. Chen, *Electron. Lett.* **34**, 7 (1998), 691–692.

Petalas, S., S. Logothetidis, S. Boultadakis, M. Alouani, and J.M. Wills, *Phys. Rev. B* **52**, 11 (1995), 8082–8091.

Polian, A., M. Grimsditch, and I. Grzegory, *J. Appl. Phys.* **79**, 6 (1996), 3343–3344.

Porowski, S., *Mater. Sci. Eng.* **B44** (1997), 407–413.

Ramirez-Flores, G., H. Navarro-Contreras, A. Lastras-Martinez, R.C. Powell, and J.E. Greene, *Phys. Rev. B* **50**, 12 (1994), 8433–8438.

Rubin, M., N. Newman, J.S. Chan, T.C. Fu, and J.T. Ross, *Appl. Phys. Lett.* **64**, 1 (1994), 64–66.

Shur, M.S., B. Gelmont, and A. Khan, *J. Electronic Mater.* **25**, 777–785 (1996).

Sichel, E.K., and Pankove J.I., *J. Phys. Chem. Solids* **38**, 3 (1977), 330.

Siegle, H., G. Kaczmarczyk, L. Filippidis, L. Litvinchuk, A. Hoffmann, and C. Thomsen, *Phys. Rev. B* **55**, 11 (1997), 7000–7004.

Strite, S., and H. Morkoc, *J. Vac. Sci. Technol. B* **10**, 4 (1992), 1237–1266.

Suzuki, M., T. Uenoyama, and A. Yanase, *Phys. Rev. B* **52**, 11 (1995), 8132–8139.

Truell, R., C. Elbaum, and B.B. Chick, *Ultrasonic Methods in Solid State Physics*, Academic Press, New York, 1969.

Wright, A.F., *J. Appl. Phys.* **82**, 6 (1997), 2833–2839.

Yu, G., G. Wang, H. Ishikawa, M. Umeno, T. Soga, T. Egawa, J. Watanabe, and T. Jimbo, *Appl. Phys. Lett.* **70**, 24 (1997), 3209–3211.

Zi, J., X. Wan, G. Wei, K. Zhang, and X. Xie, *J. Phys. Condens. Matter* **8** (1996), 6323–6328.

Aluminum Nitride (AlN)

Yu. Goldberg
The Ioffe Institute, St. Petersburg, Russia

The experimental data for cubic polytype of AlN are practically absent. In this chapter, only data for hexagonal AlN polytype have been reported.

2.1. BASIC PARAMETERS AT 300 K

Crystal structure	Wurtzite
Space group	$C_{6v}^4 - P6_3mc$
Number of atoms in 1 cm^3	9.58×10^{22}
Debye temperature (K)	1150
Density (g cm^{-3})	3.23
Dielectric constant	
static	8.5
high frequency	4.6
Effective electron mass (in units of m_0):	0.4

(Continued)

Properties of Advanced Semiconductor Materials, Edited by Levinshtein, Rumyantsev, Shur.
ISBN 0-471-35827-4   © 2001 John Wiley & Sons, Inc.

	Wurtzite
Effective hole mass (in units of m_0):	
heavy	
for k_z direction $\quad m_{hz}$	3.53
for k_x direction $\quad m_{hx}$	10.42
light	
for k_z direction $\quad m_{lz}$	3.53
for k_x direction $\quad m_{lx}$	0.24
split-off band	
for k_z direction $\quad m_{soz}$	0.25
for k_x direction $\quad m_{sox}$	3.81
Electron affinity (eV)	0.6
Lattice constant (Å)	$a = 3.112$ $c = 4.982$
Optical phonon energy (meV)	99

Band structure and carrier concentration

	Wurtzite
Energy gap (eV)	6.2
Energy separation between	
Γ valley and (M–L) valleys (eV)	$\sim$0.7
M–L-valleys degeneracy	6
Energy separation between	
Γ valley and K valleys (eV)	$\sim$1.0
K-valleys degeneracy	2
Energy of spin–orbital splitting E_{so}(eV)	0.019
Effective conduction band density of states (cm^{-3})	6.3×10^{18}
Effective valence band density of states (cm^{-3})	4.8×10^{20}

Electrical properties

	Wurtzite
Breakdown field $(V\ cm^{-1})$	$(1.2 \div 1.8) \times 10^6$
Mobility $(cm^2\ V^{-1}\ s^{-1})$	
electrons	300
holes	14
Diffusion coefficient $(cm^2\ s^{-1})$	
electrons	7
holes	0.3
Electron thermal velocity $(m\ s^{-1})$	1.85×10^5
Hole thermal velocity $(m\ s^{-1})$	0.41×10^5

(Continued)

	Wurtzite
Optical properties	
Infrared refractive index	2.15
Radiative recombination coefficient (cm^3 s^{-1})	0.4×10^{-10}
Thermal and mechanical properties	
Bulk modulus (dyn cm^{-2})	21×10^{11}
Melting point (°C)	2750 (between 100 and 500 atm of nitrogen)
Specific heat (J g^{-1} °C^{-1})	0.6
Thermal conductivity (W cm^{-1} °C^{-1})	2.85
Thermal diffusivity (cm^2 s^{-1})	1.47
Thermal expansion, linear (°C^{-1})	$\alpha_a = 4.2 \times 10^{-6}$
	$\alpha_c = 5.3 \times 10^{-6}$

2.2. BAND STRUCTURE AND CARRIER CONCENTRATION

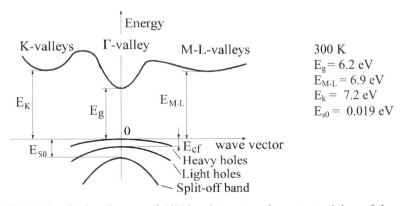

FIG. 2.2.1. Qualitative diagram of AlN band structure. Important minima of the conduction band and maxima of the valence band.

2.2.1. Temperature Dependences

Temperature dependence of energy gap

$$E_g = E_g(0) - 1.799 \times 10^{-3} \cdot \frac{T^2}{T + 1462} \,(\text{eV}) \qquad (2.2.1)$$

where T is temperature in degrees K $(0 < T < 300)$ [Guo and Yoshida (1994)].

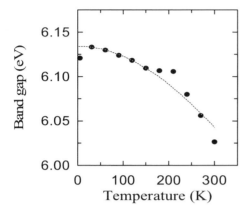

FIG. 2.2.2. Optical band gap of AlN as a function of temperature [Guo and Yoshida (1994)].

Effective density of states in the conduction band N_c:

$$N_c \cong 4.82 \times 10^{15} \cdot \left(\frac{m_\Gamma}{m_0}\right)^{3/2} \cdot T^{3/2} \; (\text{cm}^{-3})$$

$$\cong 1.2 \times 10^{15} \times T^{3/2} \; (\text{cm}^{-3}) \tag{2.2.2}$$

Effective density of states in the valence band N_v:

$$N_v = 9.4 \times 10^{16} \times T^{3/2} \; (\text{cm}^{-3}) \tag{2.2.3}$$

2.2.2. Dependences on Hydrostatic Pressure

Hydrostatic pressure dependence of the energy gap [Gorczyca and Christensen (1993)]:

$$dE_g/dP = 3.6 \times 10^{-3} \; \text{eV kbar}^{-1}$$

Conduction band first- and second-order pressure derivatives [Van Camp et al. (1991)]:

$$E_g = E_g(0) + 3.6 \times 10^{-3}P - 1.7 \times 10^{-6}P^2 \; (\text{eV}) \tag{2.2.4}$$

$$E_L = E_L(0) + 8.0 \times 10^{-4}P + 6.9 \times 10^{-7}P^2 \; (\text{eV}) \tag{2.2.5}$$

$$E_M = E_M(0) + 7.5 \times 10^{-4}P + 1.0 \times 10^{-6}P^2 \; (\text{eV}) \tag{2.2.6}$$

$$E_k = E_K(0) - 6.3 \times 10^{-4}P + 1.7 \times 10^{-6}P^2 \text{ (eV)} \qquad (2.2.7)$$

where P is pressure in kbar.

Phase transition from the wurtzite phase to the rocksalt structure (space group O_h^5; lattice parameter 4.04 Å) takes place at the pressure of 17 GPa ($\cong$173 kbar) [Gorczyca and Christensen (1993)].

2.2.3. Band Discontinuities at Heterointerfaces

GaN/AlN (0001) [Martin et al. (1996)]
    Conduction band discontinuity $\Delta E_c = 2.0$ eV
    Valence band discontinuity $\Delta E_v = 0.7$ eV

InN/AlN (0001) [Martin et al. (1996)]
    Conduction band discontinuity $\Delta E_c = 2.7$ eV
    Valence band discontinuity $\Delta E_v = 1.8$ eV

SiC/AlN (0001) [King et al. (1996)]
    Valence band discontinuity $\Delta E_v = 1.4$ eV

2.2.4. Effective Masses

Electrons

Effective mass of density of states for Γ valley: $m_\Gamma = 0.4m_0$
Theoretical estimations of the electron effective mass anisotropy in Γ valley may be found in Xu and Ching (1993)

Holes [Suzuki and Uenoyama (1996)]

heavy	
for k_z direction	$m_{hz} = 3.53m_0$
for k_x direction	$m_{hx} = 10.42m_0$
light	
for k_z direction	$m_{lz} = 3.53m_0$
for k_x direction	$m_{lx} = 0.24m_0$
split-off band	
for k_z direction	$m_{soz} = 0.25m_0$
for k_x direction	$m_{sox} = 3.81m_0$

Effective mass of density of states $m_v = 7.26m_0$

2.2.5. Donors and Acceptors

Chu et al. (1967), Francis and Worell (1976), Jenkins and Dow (1989), Tansley and Egan (1992), Mohammad et al. (1995), Boguslawski et al. (1996), Gorczyca et al. (1997).

Native donors: Si, Mg (ionization energy $\Delta E \approx 1$ eV)
Donors: C, Ge, Se
Acceptors: C, Hg

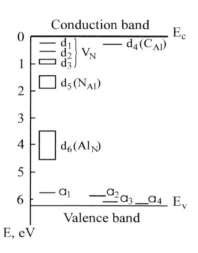

FIG. 2.2.3. Level positions in the forbidden gap of AlN [Tansley and Egan (1992)].

	Ionization Energy $(E_t - E_c)$ (eV)
d_1, d_2, d_3 are the donor levels of N vacancies (V_N):	
d_1	0.17
d_2	0.5
d_3	0.8–1.0
d_4 is the donor level of C in Al sites (C_{Al})	0.2
d_5 is the donor level of N in Al sites (N_{Al})	1.4–1.85
d_6 is the donor level of Al in N sites (Al_N)	3.4–4.5

	Ionization Energy $(E_v - E_t)$ (eV)
a_1 is the acceptor level of Al vacancies (V_{Al})	0.5
a_2 is the acceptor level of C in N sites (C_N)	0.4
a_3 is the acceptor level of Zn in Al sites (Zn_{Al})	0.2
a_4 is the acceptor level of Mg in Al sites (Mg_{Al})	0.1

2.3. ELECTRICAL PROPERTIES

2.3.1. Mobility and Hall Effect

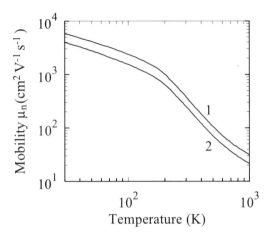

FIG. 2.3.1. The temperature dependence of phonon limited electron drift mobility calculated for two values of electron effective mass m^*. Curve 1, $m^*/m_0 = 0.42$; curve 2, $m^*/m_0 = 0.52$ [Chin et al. (1994)].

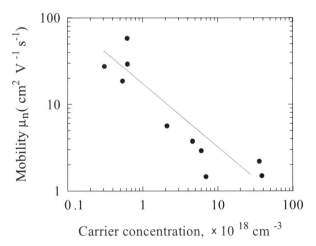

FIG. 2.3.2. The electron mobility versus electron concentration at 300 K [Wongchotiqul et al. (1996)].

Calculated electron drift mobility $\mu_n \cong 300$ cm^2 V^{-1} s^{-1} (300 K). Calculated phonon-limited electron drift mobility (for very weak doped AlN) $\mu_n \cong 2000$ cm^2 V^{-1} s^{-1} (77 K) [Chin et al. (1994)]. Hole Hall mobility $\mu_p = 14$ cm^2 V^{-1} s^{-1} (300 K) [Edwards et al. (1965)].

2.3.2. Recombination Parameters

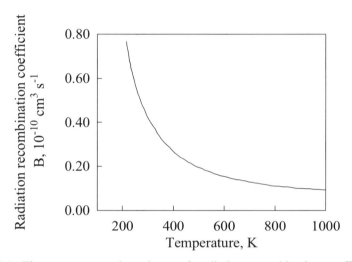

FIG. 2.3.3. The temperature dependence of radiative recombination coefficient B [Dmitriev and Oruzheinikov (1996)].

Radiative recombination coefficient at 300 K	0.4×10^{-10} cm^3 s^{-1}
Effective majority carrier (electrons) lifetime (effective lifetime of holes on the traps) [Walker et al. (1997)]	≥ 35 ms

2.4. OPTICAL PROPERTIES

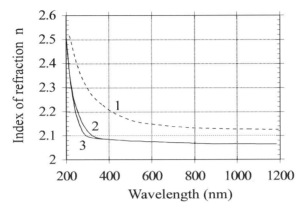

FIG. 2.4.1. Refractive index n versus wavelength. 300 K. Curve 1, Geidur and Yaskov (1980); curves 2 and 3, Demiryont et al. (1986).

Infrared Refractive Index at 300 K [Meng, (1994)]:

Epitaxial films and monocrystals	2.1–2.2
Polycrystalline films	1.9–2.1
Amorphous films	1.8–1.9

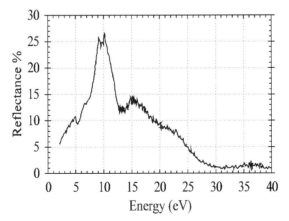

FIG. 2.4.2. Normal incidence reflectivity versus photon energy for ultraviolet region [Loughin and French (1994)].

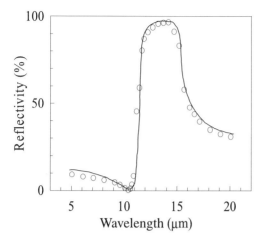

FIG. 2.4.3. Normal incidence reflectivity versus wavelength for infrared region. Circles are experimental values, solid line is calculated [Akasaki and Hashimoto (1967)]. Reprinted from *Sol. State Commun.* **5**, Akasaki, I. and Hashimoto, M., "Infrared lattice vibration of vapour-grown AlN," pp. 851–853, copyright © 1967, with permission from Elsevier Science.

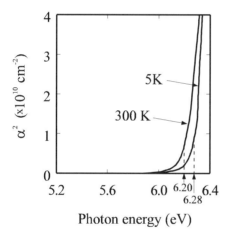

FIG. 2.4.4. The absorption coefficient squared versus photon energy near the intrinsic absorption edge for 300 K and 5 K. The values of 6.2 and 6.28 eV resulting from a straight-line fit are shown [Perry and Rutz (1978)].

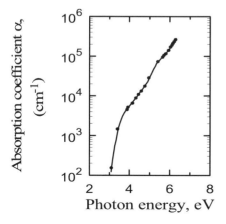

FIG. 2.4.5. The absorption coefficient versus photon energy at 300 K [Demiryont et al. (1986)].

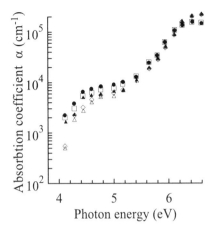

FIG. 2.4.6. The absorption coefficient versus photon energy at 300 K for sputter-deposited microcrystalline AlN grown on fused silica [Aita et al. (1989)]; see also Zarwasch et al. (1992)].

2.5. THERMAL PROPERTIES

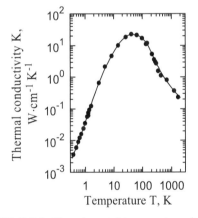

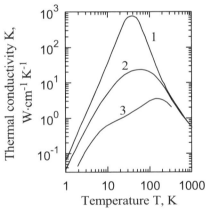

FIG. 2.5.1. Experimental temperature dependence of thermal conductivity K for high-purity AlN [Slack et al. (1987)].

FIG. 2.5.2. Temperature dependences of thermal conductivity K for AlN. Curve 1 has been estimated for a single crystal with no oxygen and a sample diameter of 0.54 cm; curve 2 was measured for a AlN single crystal containing $N_0 \cong 4.2 \times 10^{19}$ cm^{-3} oxygen atoms; curve 3, $N_0 \cong 3 \times 10^{20}$ cm^{-3} oxygen atoms [Slack et al. (1987)].

At $0.4 < T < 3$ K: $K \sim T^{2.54}$
At $500 < T < 1800$ K: $K \sim T^{-1.25}$

Maximal measured value of thermal conductivity at 300 K is 2.85 W cm^{-1} K^{-1} [Slack et al. (1987)].

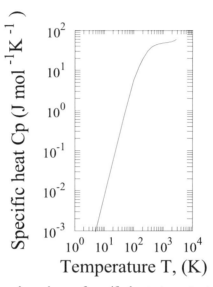

FIG. 2.5.3. Temperature dependence of specific heat at constant pressure [Koshchenko et al. (1984)].

At $300 < T < 1800$ K, $C_p = 45.94 + 3.347 \times 10^{-3} \times T - 14.98 \times 10^5 \times T^{-2}$ (J mol^{-1} K^{-1}).

At $1800 < T < 2700$ K, $C_p = 37.34 + 7.866 \times 10^{-3} \times T$ J mol^{-1} K^{-1}.

Thermal Expansion

[Slack and Bartram (1975), Touloukian at al. (1977), Meng (1994), Morkoc et al. (1994)]

For the direction along the c-axis:
At 300 K:

$$\alpha_c = 5.3 \times 10^{-6} \text{ K}^{-1}$$

At $293 < T < 1700$ K:

$$\Delta c / c_{300} = -7.006 \times 10^{-2} + 1.583 \times 10^{-4} \times T$$

$$+ 2.719 \times 10^{-7} \times T^2 - 5.834 \times 10^{-11} \times T^3$$

For the perpendicular direction:
At 300 K:

$$\alpha_a = 4.2 \times 10^{-6} \text{ K}^{-1}$$

At $293 < T < 1700$ K:

$$\Delta a / a_{300} = -8.679 \times 10^{-2} + 1.929 \times 10^{-4} \times T$$
$$+ 3.400 \times 10^{-7} \times T^2 - 7.969 \times 10^{-11} \times T^3$$

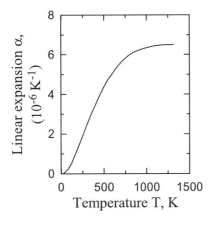

FIG. 2.5.4. Temperature dependence of linear expansion coefficient α for ceramic AlN samples:

$$\alpha = \frac{1}{3}\left(\frac{\Delta c}{c} + 2\frac{\Delta a}{a}\right)$$

[Slack and Bartram (1975)].

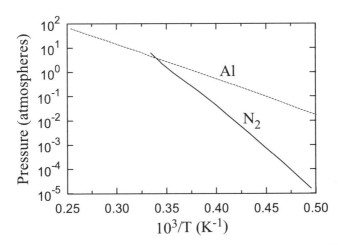

FIG. 2.5.5. Calculated vapor pressure of Al and N_2 in equilibrium with solid AlN and liquid Al [Meng (1994)].

2.6. MECHANICAL PROPERTIES, ELASTIC CONSTANTS, LATTICE VIBRATIONS, OTHER PROPERTIES

Density: 3.23 g cm^{-3}
Surface microhardness on basal plane (0001)
(using Knoop's pyramid test):   800 kg mm^{-2}
Elastic constants at 300 K
[McNeil et al. (1993)]:
    C_{11} 410 ± 10 GPa
    C_{12} 149 ± 10 GPa
    C_{13} 99 ± 4 GPa
    C_{33} 389 ± 10 GPa
    C_{44} 125 ± 5 GPa
[see also Wright (1997)]
Bulk modulus B_s (compressibility^{-1}):

$$B_s = \frac{C_{33}(C_{11}+C_{12}) - 2(C_{13})^2}{C_{11}+C_{12} - 4C_{13} + 2C_{33}}, \qquad B_s = 210 \text{ GPa}$$

Young's modulus, Y_0: $Y_0 = 308$ GPa
                                                                [Gerlich et al. (1986)]
Calculated Poisson ratio σ_0 along the different crystallographic directions [Thokala and Chaudhuri (1995)]:
{0001}, c-plane   0.287
{11$\bar{2}$0}, r-plane ($l = \langle 0001 \rangle, m = \langle 1\bar{1}00 \rangle$)   0
{11$\bar{2}$0}, r-plane ($l = \langle 1\bar{1}00 \rangle, m = \langle 0001 \rangle$)   0.216
The measured sound velocities and related elastic module deduced from them [Gerlich et al. (1986)]:
The velocity of the longitudinal waves, v_L 10,127 m s^{-1}
The velocity of the shear waves, v_s 6,333 m s^{-1}
The longitudinal elastic modulus, C_L 334 GPa
The shear elastic modulus, C_s 131 GPa
Temperature derivatives of the elastic module:

$$\frac{\partial \ln C_L}{\partial T} = -0.37 \times 10^{-4} \text{ K}^{-1}; \qquad \frac{\partial \ln C_s}{\partial T} = -0.57 \times 10^{-4} \text{ K}^{-1};$$

$$\frac{\partial \ln B_s}{\partial T} = -0.43 \times 10^{-4} \text{ K}^{-1}$$

Acoustic Wave Speeds

Wave Propagation Direction	Wave Character	Expression for Wave Speed	Wave Speed (in units of 10^5 cm/s)
[001]	V_L, longitudinal	$(C_{33}/\rho)^{1/2}$	10.97
	V_T, transverse	$(C_{44}/\rho)^{1/2}$	6.22
	V_L, longitudinal	$(C_{11}/\rho)^{1/2}$	11.27
[100]	V_T, transverse, polarization along [001]	$(C_{44}/\rho)^{1/2}$	6.22
	V_T, transverse, polarization along [010]	$[(C_{11} - C_{12})/2\rho]^{1/2}$	6.36

For definitions of the crystallographic directions see Appendix 3.

Phonon Frequencies (in cm^{-1}) [Collins et al. (1967), MacMillan et al. (1993), Perlin et al. (1993), Meng (1994)]

Mode	The Diapason of the Values Observed (cm^{-1})
$n_{TO}(E_1)$	657–673
$n_{TO}(A_1)$	607–614 (or 659–667)
$n_{LO}(E_1)$	895–924
$n_{LO}(A_1)$	888–910
$n^{(1)}(E_2)$	241–252
$n^{(2)}(E_2)$	655–660

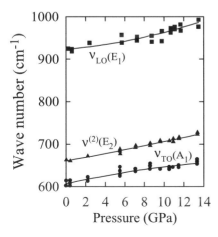

FIG. 2.6.1. Pressure dependences of the phonon frequencies (Perlin et al. [1993]).

Piezoelectric Constants
[Xinjiao et al. (1986)]

e_{15}	-0.48 C m^{-2}
e_{31}	-0.58 C m^{-2}
e_{33}	1.55 C m^{-2}

REFERENCES

Aita, C.R., C.J.G. Kubiak, and F.Y.H. Shih, *J. Appl. Phys.* **66**, 9 (1989), 4360–4367.

Akasaki, I., and M. Hashimoto, *Sol. State Commun.* **5**, 11 (1967), 851–853.

Boguslawski, P., E.L. Briggs, and J. Bernholc, *Appl. Phys. Lett.* **69**, 2 (1996), 233–235.

Chin, V.W.L., T.L. Tansley, and T. Osotchan, *J. Appl. Phys.* **75**, 11 (1994), 7365–7372.

Chu, T.L., D.W. Ing, and A.J. Noreika, *Sol. State Electron.* **10**, 10 (1967), 1023–1027.

Collins, A.T., E.C. Lightowlers, and P.J. Dean, *Phys Rev.* **158**, 3 (1967), 833–838.

Demiryont, H., L.R. Thompson, and G.J. Collins, *Appl. Optics* **25**, 8 (1986), 1311–1318.

Dmitriev, A.V., and A.L. Oruzheinikov, in *III-Nitride, SiC, and Diamond Materials for Electronic Devices* (edited, by Gaskill D.K., C.D. Brandt, and R.J. Nemanich), *Material Research Society Symposium Proceedings*, Vol. 423 (1996), pp. 69–73. Pittsburgh, PA.

Edwards, J., K. Kawabe, G. Stevens, and R.H. Tredgold, *Sol. State Commun.* **3**, (1965), 99–100.

Francis, R.W., and W.L. Worrell, *J. Electrochem. Soc.* **123**, 3 (1976), 430–433.

Geidur, S.A., and A.D. Yaskov, *Opt. Spectrosc.* **48**, 6 (1980), 618–622.

Gerlich, D., S.L. Dole, and G.A. Slack, *J. Phys. Chem. Solids* **47**, 5 (1986), 437–441.

Gorczyca, I., and N.E. Christensen, *Phys. B* **185** (1993), 410–414.

Gorczyca, I., A. Svane, and N.E. Christensen, *Internet J. Nitride Sem. Res.* **2**, Article 18 (1997).

Guo, Q., and A. Yoshida, *Jpn. J. Appl. Phys.* **33**, part 1, 5A (1994), 2453–2456.

Jenkins, D.W., and J.D. Dow, *Phys Rev. B* **39**, 5 (1989), 3317–3329.

King, S.W., M.C. Benjamin, R.J. Nemanich, R.F. Davis, and W.R.L. Lambrecht, in *Gallium Nitride and Related Materials* (edited by Ponce F.A., R.D. Dupuis, S. Nakamura, and J.A. Edmond), *Material Research Society Symposium Proceedings*, Vol. 395 (1996), pp. 375–380. Pittsburgh, PA.

Koshchenko, V.I., Ya.Kh. Grinberg, and A.F. Demidenko, *Inorg. Mater.* **20**, 11 (1984), 1550–1553.

Loughin, S., and R.H. French, in *Properties of Group III Nitrides* (edited by Edgar J.H.), EMIS Datareviews Series No. 11, 1994, an INSPEC publication, pp. 175–188.

MacMillan, M.F., R.P. Devaty, and W.J. Choyke, *Appl. Phys. Lett.* **62**, 7 (1993), 750–752.

Martin, G., A. Botchkarev, A. Rockett, and H. Morkoc, *Appl. Phys. Lett.* **68**, 18 (1996), 2541–2543.

Meng, W.J., in *Properties of Group III Nitrides* (edited by Edgar J.H.), EMIS Datareviews Series, No. 11, 1994, an INSPEC publication, pp. 22–29.

McNeil, L.E., M. Grimsditch, and R.H. French. *J. Am. Ceram. Soc.* **76**, 5 (1993), 1132–1136.

Mohammad, S.N., A.A. Salvador, and H. Morkoc, *Proc. IEEE*, **83**, 10 (1995), 1306–1355.

Morkoc, H., S. Strite, G.B. Gao, M.E. Lin, B. Sverdlov, and M. Burns, *J. App. Phys.* **76**, 3 (1994), 1363–1398.

Perlin, P., A. Polian, and T. Suski, *Phys Rev. B* **47**, 5 (1993), 2874–2877.

Perry, P.B., and R.F. Rutz, *Appl. Phys. Lett.* **33**, 4 (1978), 319–321.

Slack, G.A., and S.F. Bartram, *J. Appl. Phys.* **46**, 1 (1975), 89–98.

Slack, G.A., R.A. Tanzilli, R.O. Pohl, and J.W. Vandersande, *J. Phys. Chem. Solids* **48**, 7 (1987), 641–647.

Suzuki, M., and T. Uenoyama, *J. Appl. Phys.* **80**, 12 (1996), 6868–6874.

Tansley, T.L., and R.J. Egan, *Phys. Rev. B* **45**, 19 (1992), 10942–10950.

Thokala, R., and J. Chaudhuri, *Thin Solid Films* **266**, 2 (1995), 189–191.

Touloukian, Y.S., R.K. Kirby, R.E. Taylor, and T.Y.R. Lee (eds.) *Thermophysical Properties of Matter*, Vol. 13, Plenum Press, New York, 1977.

Van Camp, P.E., V.E. Van Doren, and J.T. Devreese, *Phys Rev. B* **44**, 16 (1991), 9056–9059.

Walker, D., X. Zhang, A. Saxler, P. Kung, J. Xu, and M. Razeghi, *Appl. Phys. Lett*, **70**, 8 (1997), 949–951.

Wongchotiqul, K., N. Chen, D.P. Zhang, X. Tang, and M.G. Spencer, in *Gallium Nitride and Related Materials* (edited by Ponce F.A., R.D. Dupuis, S. Nakamura, and J.A. Edmond), *Material Research Society Symposium Proceedings*, Vol. 395 (1996), pp. 279–282.

Wright, A.F., *J. Appl. Phys.* **82**, 6 (1997), 2833–2839.

Xinjiao, Li, Xu Zechuan, He Ziyou, Cao Huazhe, Su Wuda, Chen Zhongcai, Zhou Feng, and Wang Enguang, *Thin Solid Films* **139**, 3 (1986), 261–274.

Xu, Y.N., and W.Y. Ching, *Phys Rev. B* **48**, 7 (1993), 4335–4351.

Zarwasch, R., E. Rille, and H.K. Pulker, *J. Appl. Phys.* **71**, 10 (1992), 5275–5277.

Indium Nitride (InN)

A. Zubrilov

The Ioffe Institute, St. Petersburg, Russia

The experimental data for cubic polytype of InN are practically absent. In this chapter, only data for hexagonal (wurtzite) InN polytype have been reported

3.1. BASIC PARAMETERS AT 300 K

Crystal structure	Wurtzite
Space group	$C_{6v}^4 P6_3mc$
Number of atoms in 1 cm^3	6.4×10^{22}
Debye temperature (K)	660
Density (g cm^{-3})	6.81
Dielectric constant	
static	15.3
high frequency	8.4
Effective electron mass (in units of m_0)	0.11

(Continued)

Properties of Advanced Semiconductor Materials, Edited by Levinshtein, Rumyantsev, Shur.
ISBN 0-471-35827-4 © 2001 John Wiley & Sons, Inc.

	Wurtzite
Effective hole mass (in units of m_0)	
heavy	1.63
light	0.27
split-off band	0.65
Lattice constants (Å)	$a = 3.533$
	$c = 5.693$
Optical phonon energy (meV)	73

Band structure and carrier concentration

Energy gap (eV)	1.9–2.05

Conduction band

Energy separation between	
Γ valley and A valley (eV)	$0.7 \div 2.7$
A-valley degeneracy	1
Energy separation between	
Γ valley and Γ_1 valley (eV)	$1.1 \div 2.6$
Γ_1 valley degeneracy	1
Energy separation between	
Γ valley and M–L valleys (eV)	$2.9 \div 3.9$
M–L-valleys degeneracy	6

Valence band

Energy of spin–orbital splitting E_{so} (eV)	0.003
Energy of crystal-field splitting E_{cr} (eV)	0.017
Effective conduction band density of states (cm^{-3})	9×10^{17}
Effective valence band density of states (cm^{-3})	5.3×10^{19}

Electrical properties

Mobility ($cm^2\ V^{-1} \cdot s^{-1}$) electrons	≤ 3200
Diffusion coefficient ($cm^2 \cdot s^{-1}$) electrons	< 80
Electron thermal velocity ($m \cdot s^{-1}$)	3.4×10^5
Hole thermal velocity ($m \cdot s^{-1}$)	9.0×10^4

Optical properties

| Infrared refractive index | 2.9 |
| Radiative recombination coefficient ($cm^3\ s^{-1}$) | 2×10^{-10} |

(Continued)

	Wurtzite
Thermal and mechanical properties	
Bulk modulus (GPa)	140
Melting point (°C)	see Fig. 3.5.3
Specific heat (J g^{-1} °C^{-1})	0.32
Thermal conductivity (W cm^{-1} °C^{-1})	0.45
Thermal diffusivity (cm^2 s^{-1})	0.2
Thermal expansion, linear, (°C^{-1})	$\alpha_a = 3.8 \times 10^{-6}$ $\alpha_c = 2.9 \times 10^{-6}$

3.2. BAND STRUCTURE AND CARRIER CONCENTRATION

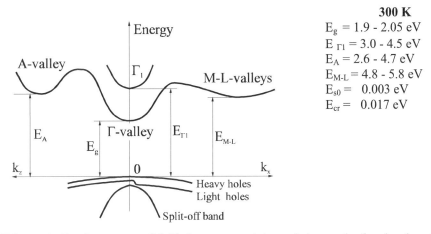

300 K
$E_g = 1.9 - 2.05$ eV
$E_{\Gamma 1} = 3.0 - 4.5$ eV
$E_A = 2.6 - 4.7$ eV
$E_{M-L} = 4.8 - 5.8$ eV
$E_{s0} = 0.003$ eV
$E_{cr} = 0.017$ eV

FIG. 3.2.1. Band structure of InN. Important minima of the conduction band and maxima of the valence band. For details see Christensen and Gorczyca (1994), Jenkins (1994), Yeo et al. (1998), and Pugh et al. (1999)].

3.2.1. Temperature Dependences

The temperature dependence of an energy gap [Guo and Yoshida (1994)] is shown in Eqs. (3.2.1) and (3.2.2).

Varshni expression:

$$E_g = E_g(0) - 2.45 \times 10^{-4} \times \frac{T^2}{T + 624} \text{ (eV)} \qquad (3.2.1)$$

Bose–Einstein expression:

$$E_g = E_g(0) - 4.39 \times 10^{-2} \times \frac{2}{\exp(466/T) - 1} \ (\text{eV}) \qquad (3.2.2)$$

$0 < T < 300$ K, where T is temperature in degrees K. See also Osamura et al. (1975).

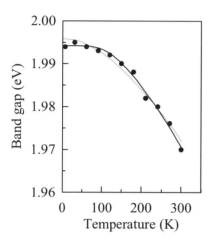

FIG. **3.2.2.** The temperature dependences of InN band gap. Broken line represents approximation (3.2.1) with $E_g(0) = 1.996$ eV. Solid line represents approximation (3.2.2) with $E_g(0) = 1.994$ eV [Guo and Yoshida (1994)].

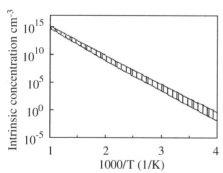

FIG. **3.2.3.** Temperature dependence of the intrinsic carrier concentration calculated for E_g magnitudes interval $1.9 \div 2.05$ eV.

Intrinsic carrier concentration:

$$n_i = (N_c \cdot N_v)^{1/2} \exp\left(-\frac{E_g}{2k_B T}\right) \qquad (3.2.3)$$

Effective density of states in the conduction band N_c:

$$N_c \cong 4.82 \times 10^{15} \cdot \left(\frac{m_\Gamma}{m_0}\right)^{3/2} \cdot T^{3/2} \ (\text{cm}^{-3})$$

$$\cong 1.76 \times 10^{14} \times T^{3/2} \ (\text{cm}^{-3}) \qquad (3.2.4)$$

Effective density of states in the valence band N_v:

$$N_v = 10^{16} \times T^{3/2} \ (\text{cm}^{-3}) \tag{3.2.5}$$

3.2.2. Dependence on Hydrostatic Pressure

For wurtzite InN:

$$E_g = E_g(0) + 3.3 \times 10^{-2}P \ (\text{eV}) \tag{3.2.6}$$

where P is pressure in GPa [Christensen and Gorczyca (1994), Perlin et al. (1997)].

3.2.3. Band Discontinuities at Heterointerfaces

InN/AlN (0001) [Martin et al. (1996)]; see also Wei and Zunger (1996)]

Conduction band discontinuity	$\Delta E_c = 2.7$ eV
Valence band discontinuity	$\Delta E_v = 1.8$ eV

InN/GaN [Martin et al. (1996)]

Conduction band discontinuity	$\Delta E_c = 0.45$ eV
Valence band discontinuity	$\Delta E_v = 1.05$ eV

3.2.4. Effective Masses

Electrons

The surfaces of equal energy in Γ valley should be ellipsoids, but effective masses in z-direction and perpendicular directions are estimated to be approximately the same.

$$m_\Gamma = 0.11m_0$$

[Lambrecht and Segall (1993)].

Holes

Heavy holes:	$m_{hh} = 1.63m_0$
Light holes:	$m_{lh} = 0.27m_0$
Split-off holes:	$m_{sh} = 0.65m_0$
Effective mass of density of state m_v:	$1.65m_0$

[Xu and Ching (1993), Yeo et al. (1998), Pugh et al. (1999)]

3.2.5. Donors and Acceptors

Ionization energy of shallow donor [native defect level (V_N)]:
<40–50 meV
[Tansley and Egan (1992); see also Jenkins and Dow (1989)].

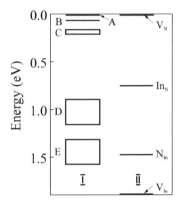

FIG. 3.2.4. Level positions in the forbidden gap of InN. Column 1 shows experimental data that fall into five groups A–E. Column 2 are the calculated energies of point defects [Tansley and Egan (1992)].

3.3. ELECTRICAL PROPERTIES

3.3.1. Mobility and Hall Effect

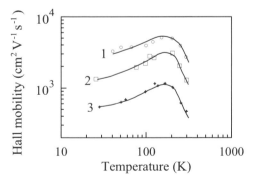

FIG. 3.3.1. Electron Hall mobility versus temperature for different doping levels and different degrees of compensation θ of polycrystalline InN films. 1—$N_d = 5.1 \times 10^{16}$ cm^{-3}, $\theta \approx 0.3$; 2—$N_d = 8.7 \times 10^{16}$ cm^{-3}, $\theta \approx 0.6$; 3—$N_d = 3.9 \times 10^{17}$ cm^{-3}, $\theta \approx 0.8$ [Tansley and Foley (1984)].

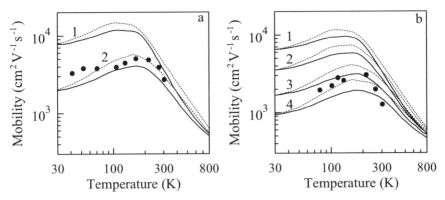

FIG. 3.3.2. The temperature dependences of electron drift (solid curves) and Hall (dashed curves) mobility of InN calculated for different carrier concentrations at different compensation ratios θ. (a) $n_0 = 5 \times 10^{16}$ cm^{-3}. 1, $\theta = 0$; 2, $\theta = 0.60$. (b) $n_0 = 8 \times 10^{16}$ cm^{-3}. 1, $\theta = 0$, 2, $\theta = 0.30$; 3, $\theta = 0.60$; 4, $\theta = 0.75$. Experimental data are taken from Tansley and Foley (1984) [Chin et al. (1994)].

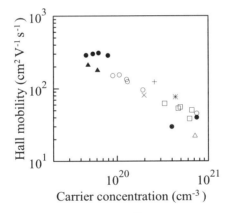

FIG. 3.3.3. The electron Hall mobility of InN films grown on sapphire and GaAs substrates as a function of carrier concentration [Yamamoto et al. (1998)].

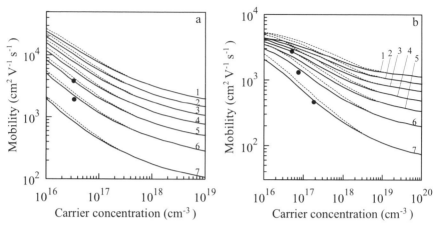

FIG. 3.3.4. The electron drift (solid curves) and Hall (dashed curves) mobility of InN calculated as a function of carrier concentration at different compensation ratios θ. (a) $T = 77$ K. (b) $T = 300$ K. 1, $\theta = 0$; 2, $\theta = 0.15$; 3, $\theta = 0.30$; 4, $\theta = 0.45$; 5, $\theta = 0.60$; 6, $\theta = 0.75$; 7, $\theta = 0.90$. Experimental data are taken from Tansley and Foley (1984) [Chin et al. (1994)].

3.3.2. Transport Properties in High Electric Field

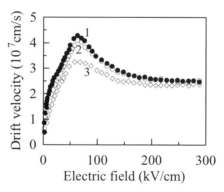

FIG. 3.3.5. Calculated steady-state drift velocity of InN as a function of electric field at different doping concentrations N_d. $T = 300$ K. N_d (cm^{-3}): 1, 10^{17}; 2, 10^{18}; 3, 10^{18} [O'Leary et al. (1998)].

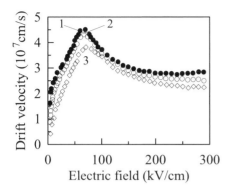

FIG. 3.3.6. Calculated steady-state drift velocity of InN as a function of electric field at different temperatures T. Doping concentration $N_d = 10^{17}$ cm^{-3}. T (K): 1, 150; 2, 300; 3, 500 [O'Leary et al. (1998)].

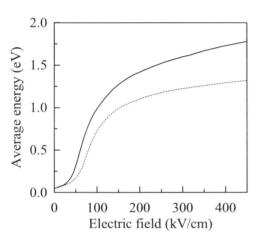

FIG. 3.3.7. Calculated average electron energy as a function of electric field. The continuous and dashed lines correspond to applied electric field along Γ–M and Γ–A directions, respectively [Bellotti et al. (1999)].

3.3.3. Impact Ionization

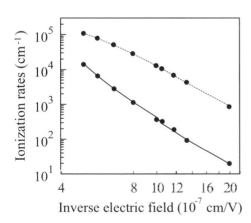

FIG. 3.3.8. Calculated electron impact ionization rates as a function of inverse electric field at 300 K. The continuous and dashed lines correspond to applied electric field along Γ–M and Γ–A directions, respectively [Bellotti et al. (1999)].

3.3.4. Recombination Parameters

Only calculated data are available for the radiative recombination coefficient in InN. The radiative recombination coefficient is 2×10^{-10} cm^3 s^{-1}
at 300 K [Zhou et al. (1995)].

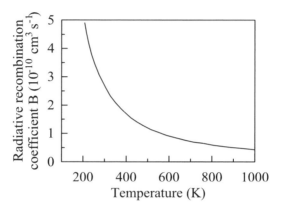

FIG. 3.3.9. The temperature dependence of radiative recombination coefficient B
[Dmitriev and Oruzheinikov (1996)].

3.4. OPTICAL PROPERTIES

Infrared refractive index (300 K): $\qquad n_\infty = (k_\infty)^{1/2} \approx 2.9$

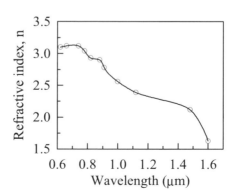

FIG. 3.4.1. Refractive index n versus wavelength of InN at 300 K [Tyagai et al. (1977)].

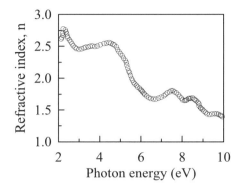

FIG. 3.4.2. Refractive index n versus photon energy of InN at 300 K [Djurisic and Li (1999)].

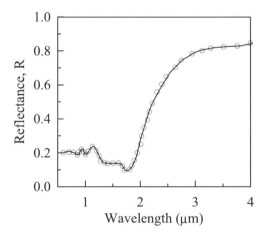

FIG. 3.4.3. Reflectance R as a function of wavelength of InN polycrystalline film at 300 K [Tyagai et al. (1977)].

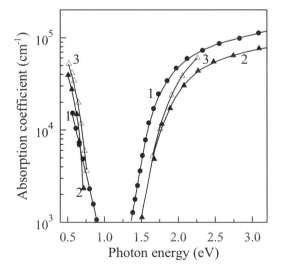

FIG. 3.4.4. The absorption coefficient versus photon energy for InN samples with free electron concentration (cm^{-3}): 5×10^{18} (curve 1); $3–6 \times 10^{20}$ (curves 2 and 3). Curves 1 and 2, $T = 300°C$; curve 3, $T = 150°C$ [Trainor and Rose (1974)].

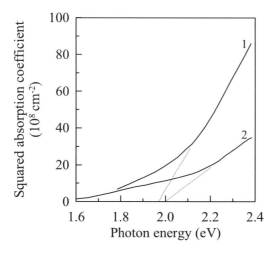

FIG. 3.4.5. The squared absorption coefficient as a function of photon energy at $T = 300$ K (curve 1) and $T = 4.2$ K (curve 2) [Puychevrier and Menoret (1976)].

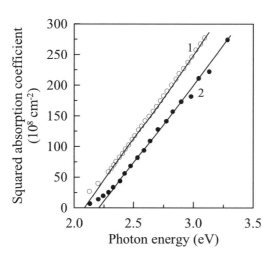

FIG. 3.4.6. The squared absorption coefficient as a function of photon energy at $T = 300$ K (curve 1) and $T = 4.2$ K (curve 2) [Guo and Yoshida (1994)].

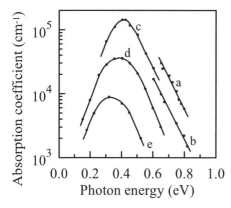

FIG. 3.4.7. Infrared absorption coefficient as a function of photon energy for the samples with different concentrations n. n (cm^{-3}): a, 3×10^{20}; b, 5×10^{18}; c, 10^{19}; d, 10^{18}; e, 5×10^{16} [Tansley and Foley (1986)].

3.5. THERMAL PROPERTIES

Thermal conductivity at 300 K:

0.45 W cm^{-1} °C^{-1} experiment [Krukowski et al. (1998)]
1.76 W cm^{-1} °C^{-1} estimate for ideal InN crystal
                    [Krukowski et al. (1998)]

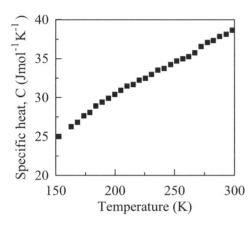

FIG. 3.5.1. The temperature dependence of specific heat [Krukowski et al. (1998)].

The specific heat C_p of InN at constant pressure for $298\ K < T < 1273$ K [Barin et al. (1977)]:

$$C_p = 38.1 + 1.21 \times 10^{-2} \cdot T \qquad (\text{J mol}^{-1}\ \text{K}^{-1})$$

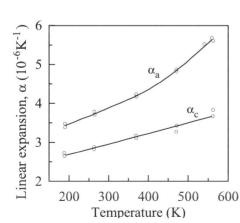

FIG. 3.5.2. Temperature dependence of linear expansion coefficients [Sheleg and Savastenko (1976)].

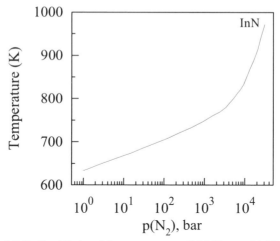

FIG. 3.5.3. Equilibrium N_2 pressure over InN [Porowski (1997)].

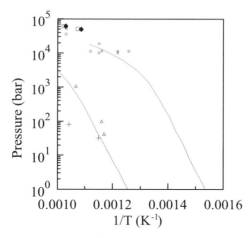

FIG. 3.5.4. Pressure–temperature stability of InN. Upper curve, Grzegory et al. (1994); lower line, McChesney et al. (1970). Symbols above the lines are stability points, symbols below the lines are instability points [Krukowski et al. (1998)]. Reprinted from *J. Phys. Chem. Solids* **59**, Krukowski, S., Witek, A., Adamczyk, J., Jun, J., Bockowski, M., Grzegory, I., Lucznik, B., Nowak, G., Wroblewski, M., Presz, A., Gierlotka, S., Stelmach, S., Palosz, B., Porowski, S., and Zinn, P., "Thermal properties of indium nitride," pp. 289–295, copyright © 1998, with permission from Elsevier Science.

3.6. MECHANICAL PROPERTIES, ELASTIC CONSTANTS, LATTICE VIBRATIONS, OTHER PROPERTIES

Density: 6.81 g cm^{-3}

Wurtzite

Elastic constants at 300 K
[Sheleg and Savastenko (1979)]:

C_{11} 190 $\pm$ 7 GPa
C_{12} 104 $\pm$ 3 GPa
C_{13} 121 $\pm$ 7 GPa
C_{33} 182 $\pm$ 6 GPa
C_{44} 10 $\pm$ 1 GPa

[see also Wright (1997) and Kim et al. (1996)]
Bulk modulus B_s (compressibility^{-1}):

$$B_s = \frac{C_{33}(C_{11} + C_{12}) - 2(C_{13})^2}{C_{11} + C_{12} - 4C_{13} + 2C_{33}}, \qquad B_s = 140 \text{ GPa}$$

Acoustic Wave Speeds

Wave Propagation Direction	Wave Character	Expression for Wave Speed	Wave Speed (in units of 10^5 cm/s)
[001]	V_L (longitudinal)	$(C_{33}/\rho)^{1/2}$	5.17
	V_T (transverse)	$(C_{44}/\rho)^{1/2}$	1.21
[100]	V_L (longitudinal)	$(C_{11}/\rho)^{1/2}$	5.28
	V_T (transverse, polarization along [001])	$(C_{44}/\rho)^{1/2}$	1.21
	V_T (transverse, polarization along [010])	$[(C_{11} - C_{12})/2\rho]^{1/2}$	2.51

For definitions of the crystallographic directions, see Appendix 3.

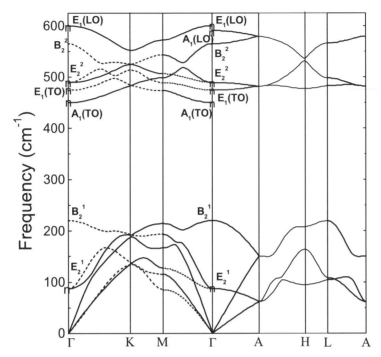

FIG. 3.6.1. Dispersion curves for acoustic and optical branch phonons for InN [Davydov et al. (1999)].

Phonon Frequencies (in cm^{-1})
[Davydov et al. (1999)]

Mode	Frequency
$E_2(\text{low})$	87
$E_2(\text{high})$	488
$A_1(\text{TO})$	451
$A_1(\text{LO})$	588
$E_1(\text{TO})$	476
$E_1(\text{LO})$	592

Piezoelectric Constants
[Bernardini and Fiorentini (1997)]

e_{31}	-0.57 C m^{-2}
e_{33}	0.97 C m^{-2}

REFERENCES

Barin, I., O. Knacke, and O. Kubaschewski, *Thermodynamical Properties of Inorganic Substances*, Springer-Verlag, Berlin, 1977.

Bellotti, E., B.K. Doshi, K.F. Brennan, J.D. Albrecht, and P.P. Ruden, *J. Appl. Phys.* **85** (1999), 916–923.

Bernardini, F., V. Fiorentini, and D. Vanderbilt, *Phys. Rev. B* **56** (1997), R10024–R10027.

Chin, V.W.L., T.L. Tansley, and T. Osotchan, *J. Appl. Phys.* **75** (1994), 7365–7372.

Christensen, N.E., and I. Gorczyca, *Phys. Rev. B* **50** (1994), 4397–4415.

Davydov, V.Yu., V.V. Emtsev, I.N. Goncharuk, A.N. Smirnov, V.D. Petrikov, V.V. Mamutin, V.A. Vekshin, S.V. Ivanov, M.B. Smirnov, and T. Inushima, *Appl. Phys. Lett* **75** (1999), 3297–3299.

Djurisic, A.B., and E.H. Li, *J. Appl. Phys.* **85** (1999), 2848–2853.

Dmitriev, A.V., and A.L. Oruzheinikov, *Mat. Res. Soc. Symp.* **423** (1996), 69–70.

Grzegory, I., S. Krukowski, J. Jun, M. Bockowski, M. Wroblewski, and S. Porowski, *AIP Conf. Proc.* **309** (1994), 565.

Guo, Q., and A. Yoshida, *Jpn. J. Appl. Phys.* **33** (1994), 2453–2456.

Jenkins, D., Band structure of InN, GaInN and AlInN, in *Properties of Group III Nitrides*, edited by Edgar, J., EMIS Datareviews Series, No. 11, 1994, pp. 159–166.

Jenkins, D.W., and J.D. Dow, *Phys. Rev. B* **39** (1989), 3317–3329.

Kim, K., W.R.L. Lambrecht, and B. Segal, *Phys. Rev. B* **53** (1996), 16310–16326.

Krukowski, S., A. Witek, J. Adamczyk, J. Jun, M. Bockowski, I. Grzegory, B. Lucznik, G. Nowak, M. Wroblewski, A. Presz, S. Gierlotka, S. Stelmach, B. Palosz, S. Porowski, and P. Zinn, *J. Phys. Chem. Solids* **59** (1998), 289–295.

Lambrecht, W.R., and B. Segall, *Phys. Rev. B* **47** (1993), 9289–9296.

Martin, G., A. Botchkarev, A. Rockett, and H. Morkoc, *Appl. Phys. Lett.* **68** (1996), 2541–2543.

McChesney, J.B., P.M. Brindenbaugh, and O.B. O'Connor, *Mater. Res. Bull.* **5** (1970), 783.

O'Leary, S.K., B.E. Foutz, M.S. Shur, U.V. Bhapkar, and L.F. Eastman, *J. Appl. Phys.* **83** (1998), 826–829.

Osamura, K., S. Naka, Y. Murakami, *J. Appl. Phys.* **46** (1975), 3432–3437.

Perlin, P., V. Iota, B.A. Weinstein, P. Wisniewski, T. Suski, P.G. Eliseev, and M. Osinski, *Appl. Phys. Lett.* **70** (1997), 2993–2995.

Porowski, S., *Mater. Sci. Eng.* **B44** (1997), 407–413.

Pugh, S.K., D.J. Dugdale, S. Brand, and R.A. Abram, *Semicond. Sci. Techn.* **14** (1999), 23–31.

Puychevrier, N., and M. Minoret, *Thin Solid Films* **36** (1976), 141.

Sheleg, A.U., and V.A. Savastenko, *Vesti Akad. Nauk BSSR, Ser. Fiz. Mater. Nauk 3*, (1976), 126.

Sheleg, A.U., and V.A. Savastenko, *Izv. Akad. Nauk SSSR. Neorg. Mater.* **15** (1979), 1598.

Tansley, T.L., and R.J. Egan, *Phys. Rev. B* **45** (1992), 10942–10950.

Tansley, T.L., and C.P. Foley, *Electron. Lett.* **20** (1984), 1066–1068.

Tansley, T.L., and C.P. Foley, *J. Appl. Phys.* **59** (1986), 3241–3244.

Trainor, J.W., and K. Rose, *J. Electron. Mater.* **3** (1974), 821–828.

Tyagai, V.A., A.M. Evstigneev, A.N. Krasiko, A.F. Andreeva, V.Ya. Malakhov, *Sov. Phys.-Semicond.* **11** (1977), 1257–1259.

Wei, S.H., and A. Zunger, *Appl. Phys. Lett.* **69** (1996), 2719–2721.

Wright, A.F., *J. Appl. Phys.* **82** (1997), 2833–2839.

Xu, Y-N., and W.Y. Ching, *Phys. Rev. B* **48** (1993), 4335–4350.

Yamamoto, A., T. Shin-ya, T. Sugiura, and A. Hashimoto, *J. Cryst. Growth* **189/190** (1998), 461–465.

Yeo, Y.C., T.C. Chong, and M.F. Li, *J. Appl. Phys.* **83** (1998), 1429–1436.

Zhou, B., K.S.A. Butcher, X. Li, and T.L. Tansley, *Abstracts of Topical Workshop on III–V Nitrides*, TWN '95, Nagoya, Japan 1995, p. 7.

Chapter 4

Boron Nitride (BN)

S. Rumyantsev, M. Levinshtein
The Ioffe Institute, St. Petersburg, Russia

A.D. Jackson, S.N. Mohammad, G.L. Harris, and M.G. Spencer
Howard University, Washington, D.C.

M. Shur
Rensselaer Polytechnic Institute, Troy, New York

Boron nitride (BN) has at least four crystal modifications of wurtzite, zinc blende, hexagonal, and rhombehedral. This chapter covers the properties of the first three (more common) modifications.

Wurtzite structure (also known as γ-BN) was first synthesized in 1963. Typically, BN crystals with wurtzite symmetry are very small (fraction of microns), are highly defective, and contain other phases.

Zinc blende modification of BN (also known as cubic or sphalerite or β-BN) was first synthesized in 1957 using the technique similar to that used for diamond growth. Now crystals with a few millimeter sizes are commercially available.

Hexagonal BN (also known as α-BN) with the structure similar to graphite is known for more than a century. Many properties of hexagonal BN are highly anisotropic and depend on the growth method. In

Properties of Advanced Semiconductor Materials, Edited by Levinshtein, Rumyantsev, Shur.
ISBN 0-471-35827-4 © 2001 John Wiley & Sons, Inc.

many cases, the different values of α-BN physical parameters given in this chapter reflect the differences in material properties of hexagonal BN grown by different methods.

4.1. BASIC PARAMETERS AT 300 K

Crystal structure	Wurtzite	Zinc Blende	Hexagonal
Group of symmetry	$C_{6v}{}^4 P6_3mc$	$T_d^2 - F\bar{4}3m$	$D_{6v}P6_3mmc$
Debye temperature (K)	1400	1700	400
Density (g/cm^3)	3.48	3.450	2.0–2.28
Dielectric constant			
static	$\varepsilon_\perp = 6.8$	7.1	$\varepsilon_\perp = 6.85$
	$\varepsilon_\parallel = 5.1$		$\varepsilon_\parallel = 5.06$
high frequency	$\varepsilon_\perp \approx \varepsilon_\parallel =$	4.46	$\varepsilon_\perp = 4.3$
	4.2–4.5		
			$\varepsilon_\parallel = 2.2$
Effective electron mass (in units of m_0)			
longitudinal m_l	0.35	1.2	
transversal m_t	0.24	0.26	
in the direction M $\rightarrow$ Γ			0.26
in the direction M $\rightarrow$ L			2.21
Effective hole masses (in units of m_0)			
in the direction $\Gamma \rightarrow$ K	0.88	$m_1 \approx 3.16$	
		$m_2 \approx 0.64$	
		$m_3 \approx 0.44$	
in the direction $\Gamma \rightarrow$ A	1.08		
in the direction $\Gamma \rightarrow$ M	1.02		
in the direction $\Gamma \rightarrow$ X		0.55	
in the direction $\Gamma \rightarrow$ L		$m_1 \approx 0.36$	
		$m_2 \approx 1.20$	
in the direction $K \rightarrow \Gamma$			0.47
in the direction $M \rightarrow \Gamma$			0.50
in the direction M $\rightarrow$ L			1.33

(Continued)

	Wurtzite	Zinc Blende	Hexagonal
Lattice constants (Å)	$a = 2.55$	3.615	$a = 2.5$–2.9
	$c = 4.17$		$c = 6.66$
Optical phonon energy (meV)	~ 130	~ 130	

Band structure and carrier concentration

Energy gap (eV)	4.5–5.5	$6.1 \div 6.4$	$4.0 \div 5.8$

Conduction band

Energy separation E_Γ (eV)	8.5	8.5–10	9
Energy separation E_M (eV)	6.6		
Energy separation E_L (eV)		>12	
Energy separation E_A (eV)			10
Effective conduction band density of states (cm^{-3})	1.5×10^{19}	2.1×10^{19}	
Effective valence band density of states (cm^{-3})	2.6×10^{19}	2.6×10^{19}	

Electrical properties

Breakdown field (V cm^{-1})		$(2 \div 6) \times 10^6$	$(1 \div 3) \times 10^6$
Mobility (cm^2 V^{-1} s^{-1})			
electrons		≤ 200	
holes		≤ 500	
Diffusion coefficient (cm$^2 \cdot$ s^{-1})			
electrons		≤ 5	
holes		≤ 12	

Optical properties

Infrared refractive index	2.05	2.1	1.8

Thermal and mechanical properties

Bulk modulus (GPa)	400	400	36.5
Melting point (°C)		see Figs. 4.5.14 and 4.5.15	
Specific heat (J g^{-1} °C^{-1})	~ 0.75	~ 0.6	~ 0.8

(Continued)

	Wurtzite	Zinc Blende	Hexagonal
Thermal conductivity $(\text{W cm}^{-1}\ {}^{\circ}\text{C}^{-1})$			
experimentally achieved		7.4	
theoretically estimated		~ 13	
parallel to the c axis			≤ 0.03
perpendicular to the c axis			≤ 6
Thermal expansion, linear $({}^{\circ}\text{C}^{-1})$		1.2×10^{-6}	
parallel to the c axis	2.7×10^{-6}		38×10^{-6}
perpendicular to the c axis	2.3×10^{-6}		-2.7×10^{-6}

4.2. BAND STRUCTURE AND CARRIER CONCENTRATION

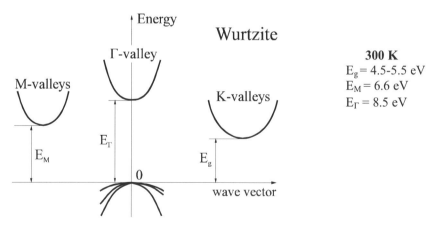

FIG. 4.2.1. Band structure of wurtzite BN. Important minima of the conduction band and maxima of the valence band. [For details see Yong-Nian Xu and Ching (1991) and Christersen and Gorczyca (1994)].

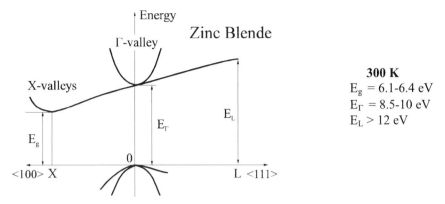

FIG. 4.2.2. Band structure of zinc blende BN. Important minima of the conduction band and maxima of the valence band. [For details see Rodriguez-Hernandez et al. (1995) and Ferhat et al. (1998)].

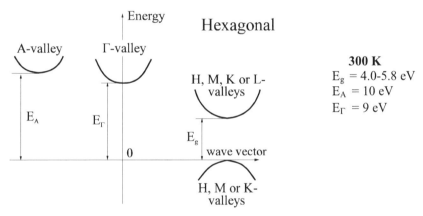

FIG. 4.2.3. Band structure of hexagonal (graphite-like) BN. Important minima of the conduction band and maxima of the valence band. The energy gaps between the top of the valence band and H, M, K, and L valleys of the conduction band are of the same order of magnitude [For details and references see Yong-Nian Xu and Ching (1991), Zunger et al. (1976), and Taylor and Clarke (1997)]. The energies of the valence band maxima are very close in the points K, H, and M of Brillouin zone.

Effective Density of States in the Conduction Band, N_c

Only the calculated data available for the values of electron effective masses for all types of BN crystals (see Section 4.2.2).

Wurtzite

$$N_c \cong 4.82 \times 10^{15} \cdot \left(\frac{m_{cd}}{m_0}\right)^{3/2} \cdot T^{3/2} \; (\text{cm}^{-3})$$

$$\cong 2.8 \times 10^{15} \times T^{3/2} \; (\text{cm}^{-3}) \qquad (4.2.1)$$

Zinc blende

$$N_c \cong 4.82 \times 10^{15} \cdot \left(\frac{m_{cd}}{m_0}\right)^{3/2} \cdot T^{3/2} \; (\text{cm}^{-3})$$

$$\cong 4.1 \times 10^{15} \times T^{3/2} \; (\text{cm}^{-3}) \qquad (4.2.2)$$

Effective Density of States in the Valence Band, N_v

Only the calculated data available for the values of hole effective masses for all types of BN crystals. These calculations are fairly inaccurate. As a crude estimate, the effective density of states in the valence band, N_v:

$$N_v \approx 5.0 \times 10^{15} \times T^{3/2} \; (\text{cm}^{-3}) \qquad (4.2.3)$$

could be used for all crystal modifications of BN (see Section 4.2.2).

4.2.1. Dependence on Hydrostatic Pressure

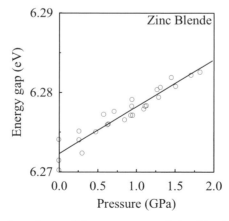

FIG. 4.2.4. Pressure dependence of the energy gap of zinc blende BN [Onodera et al. (1993)].

According to Kim et al. (1996), for zinc blende BN we have

$$dE_g/dP = 3.0 \text{ meV/GPa}$$

For wurtzite BN we have

$$dE_g/dP = 3.8 \text{ meV/GPa [Kim et al. (1996)]}$$

4.2.2. Effective Masses [Xu and Ching (1991)]

Electrons

For wurtzite BN:
The surfaces of equal energy are ellipsoids.

Longitudinal effective mass:	$m_l = 0.35m_0$
Transversal effective mass:	$m_t = 0.24m_0$

Effective mass of density of states in one valley of conduction band:

$$m_c = (m_l \times m_t^2)^{1/3} = 0.27m_0$$

Effective mass of conductivity m_{cc}:

$$m_{cc} = 3 \times \left(\frac{1}{m_l} + \frac{2}{m_t} \right)^{-1} \approx 0.27m_0$$

Effective mass of density of states for all valleys of conduction band:

$$m_{cd} \approx 0.7m_0$$

For zinc blende BN:
The surfaces of equal energy are ellipsoids.

Longitudinal effective mass:	$m_l = 1.2m_0$
Transversal effective mass:	$m_t = 0.26m_0$

Effective mass of density of states in one valley of conduction band:

$$m_c = (m_l \times m_t^2)^{1/3} = 0.43m_0$$

Effective mass of conductivity m_{cc}:

$$m_{cc} = 3 \times \left(\frac{1}{m_l} + \frac{2}{m_t}\right)^{-1} \approx 0.35m_0$$

Effective mass of density of states for all valleys of conduction band:

$$m_{cd} \approx 0.9m_0$$

For hexagonal BN:

Effective mass in the direction M → Γ: $0.26m_0$
Effective mass in the direction M → L: $2.21m_0$

Holes

For wurtzite BN:

Effective mass in the direction Γ → K: $0.88m_0$
Effective mass in the direction Γ → A: $1.08m_0$
Effective mass in the direction Γ → M: $1.02m_0$
Effective mass of density of states: $m_v \approx 1.0m_0$

For zinc blende BN:

Effective mass in the direction Γ → K: $m_1 \approx 3.16m_0$
                                                                      $m_2 \approx 0.64m_0$
                                                                      $m_3 \approx 0.44m_0$
The effective mass in the direction Γ → X: $m \approx 0.55m_0$
The effective mass in the direction Γ → L: $m_1 \approx 0.36m_0$
                                                                      $m_2 \approx 1.20m_0$

For hexagonal BN:

The effective mass in the direction K → Γ: $m \approx 0.47m_0$
The effective mass in the direction M → Γ: $m_1 \approx 0.50m_0$
The effective mass in the direction M → L: $m_1 \approx 1.33m_0$

4.2.3. Donors and Acceptors

Zinc Blende BN [Wentorf (1957)], Mishima et al. (1987), Taniguchi et al. (1993), Gubanov et al. (1997)]

Ionization energies of donors (eV):

Si	0.24
C	0.28–0.41
S	0.05

Ionization energies of acceptor (eV):

Be	0.19–0.3

Hexagonal BN [Lopatin (1994)]

Ionization energies of donors (eV)
native defects: $0.7 \div 1.5$ (group of the levels)
Ionization energies of acceptors (eV)
native defects: ≤ 1.5 (group of the levels)

For data on doping with Mg see Lu et al. (1996).

4.3. ELECTRICAL PROPERTIES

The data on the electrical properties of BN are very limited.

Zinc Blende

The majority of papers published are devoted to zinc blende modification: Bam et al. (1976), Mishima et al. (1987), Bar-Yam et al. (1992), Taniguchi et al. (1993), Lu et al. (1996), Litvinov et al. (1998).

Unintentionally doped BN films are *p*-type with carrier concentrations in the high 10^{16} to low 10^{17} cm^{-3} levels. It had been suggested that the unintentional dopants are nitrogen vacancies [Bar-Yam et al. (1992)].

The highest mobility achieved:

For n-type, $\mu_n \approx 200$ cm^2/V $\cdot$ s for $N_d = 6 \times 10^{16}$ cm^{-3} [Waters (1995)].

For p-type, $\mu_p \approx 500$ cm^2/V $\cdot$ s at room temperature for carrier concentration $p = 5 \times 10^{18}$ cm^{-3} [Litvinov et al. (1998)].

The electrical breakdown field: $(2-6) \times 10^6$ V/cm

[Brozek et al. (1994)].

Hexagonal

The electrical breakdown field: $(1-3) \times 10^6$ V/cm

[Lopatin (1994)].

4.4. OPTICAL PROPERTIES

Infrared refractive index of wurtzite BN at 300 K: $n_\infty \approx 2.05$
Infrared refractive index of zinc blende BN at 300 K: $n_\infty \approx 2.1$
Infrared refractive index of hexagonal BN at 300 K: $n_\infty \approx 1.8$

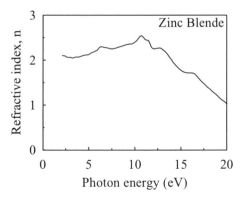

FIG. 4.4.1. Refractive index n versus photon energy of zinc blende BN [Miyata et al. (1989)].

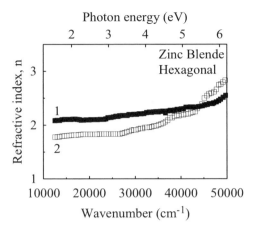

FIG. 4.4.2. Refractive index *n* versus wavenumber for zinc blende (curve 1) and hexagonal (curve 2) BN [Stenzel et al. (1996)].

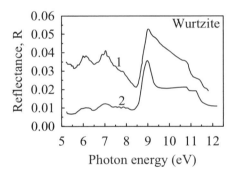

FIG. 4.4.3. Reflectance *R* as a function of photon energy of two samples of wurtzite BN. 1, 500 °C annealed; 2, 100 °C annealed (nanoscale powder compacted into dense solid under high pressure) [Yixi et al. (1994)].

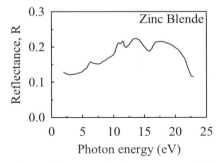

FIG. 4.4.4. Reflectance *R* as a function of photon energy of zinc blende BN [Miyata et al. (1989)].

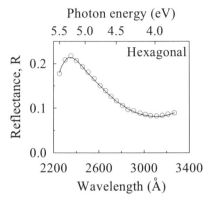

FIG. 4.4.5. Reflectance R as a function of wavelength of hexagonal BN [Zunger et al. (1976)].

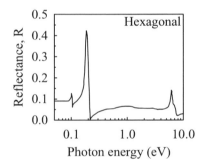

FIG. 4.4.6. Reflectance R as a function of wavelength of hexagonal BN [Hoffman et al. (1984)].

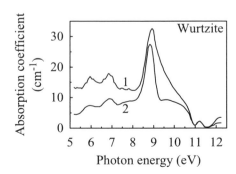

FIG. 4.4.7. The absorption coefficient as a function of photon energy for two samples of wurtzite BN. 1, 500 °C annealed; 2, 100 °C annealed. (nanoscale powder compacted into dense solid under high pressure) [Yixi et al. (1994)].

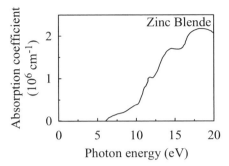

FIG. 4.4.8. The absorption coefficient as a function of photon energy of zinc blende BN [Miyata et al. (1989)].

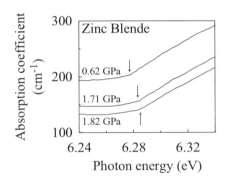

FIG. 4.4.9. The absorption coefficient versus photon energy of zinc blende BN at different hydrostatic pressures. The energies shown by arrows are defined as indirect band gaps [Onodera et al. (1993)].

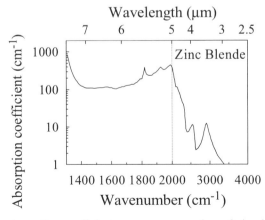

FIG. 4.4.10. The absorption coefficient versus wavenumber of zinc blende BN in the infrared [Chrenko (1974)].

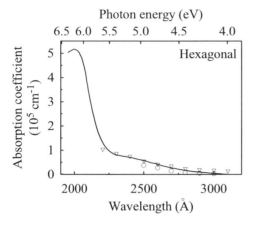

FIG. 4.4.11. The absorption coefficient versus wavelength of hexagonal BN at 300 K [Zunger et al. (1976)].

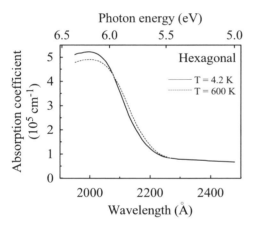

FIG. 4.4.12. The absorption coefficient versus wavelength of hexagonal BN at 4.2 K and 600 K [Zunger et al. (1976)].

4.5. THERMAL PROPERTIES

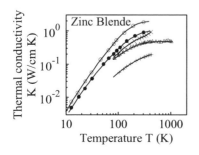

FIG. 4.5.1. Temperature dependence of thermal conductivity for different zinc blende BN samples [Slack (1973), Makedon et al. (1972)].

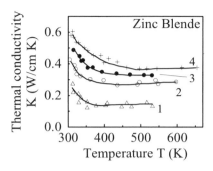

FIG. **4.5.2.** Temperature dependences of the thermal conductivity of undoped (curve 1) and Se doped (curves 2–4) zinc blende BN before and after annealing at 900–1000 K. Selenium concentration (cm^{-3}): curve 2, 2.4×10^{18}, before annealing; curve 3, 2.4×10^{18}, annealed; curve 4, 10^{19}, annealed [Shipilo et al. (1986)].

The highest thermal conductivity achieved for single-crystal zinc blende BN is 7.4 W cm^{-1} K^{-1} [Novikov et al. (1983)].

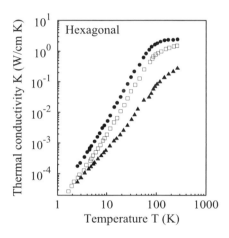

FIG. **4.5.3.** Thermal conductivity perpendicular to the c axis as a function of temperature of three hexagonal BN samples deposited at different temperatures [Duclaux et al. (1992)].

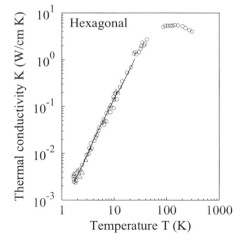

FIG. **4.5.4.** Thermal conductivity perpendicular to the c axis as a function of temperature of highly oriented hexagonal BN sample [Sichel et al. (1976)].

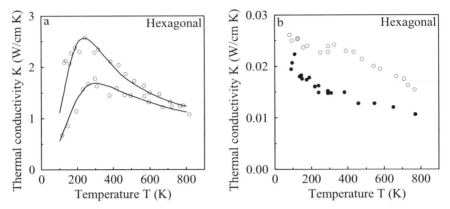

FIG. 4.5.5. Thermal conductivity of hexagonal BN perpendicular (a) and parallel (b) to the c axis as a function of temperature for two samples [Simpson and Stuckes (1971)].

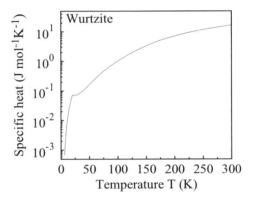

FIG. 4.5.6. Temperature dependence of the specific heat of wurtzite BN [Gorbunov et al. (1988); see also Sirota and Kofman (1976) and Inaba and Yoshiasa (1997)]. The abnormality with extremum at 21 K is caused by the presence of the ordered point defects system in the lattice [Solozhenko (1994)].

At 420 K $< T <$ 980 K, the specific heat C_p of wurtzite BN can be approximated as

$$C_p = 48.35 \times \left(\frac{T^2}{T^2 - 8.37 \times T + 68306} \right)^2 \text{ (J/Kmol)} \qquad (4.5.1)$$

[Solozhenko (1994)].

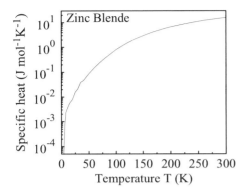

FIG. 4.5.7. Temperature dependence of the specific heat of zinc blende BN at low temperatures (single crystal) [Solozhenko et al. (1987); see also Sirota and Kofman (1975)].

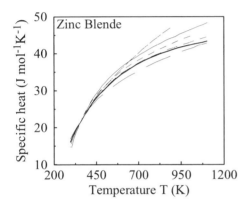

FIG. 4.5.8. Temperature dependence of high temperature specific heat of zinc blende BN according to different authors [Lyusternik and Solozhenko (1992)].

At $300 < T < 1100$ K, the specific heat C_p of zinc blende BN can be approximated as [Lyusternik and Solozhenko (1992)]

$$C_p = 48.4 \times \left(\frac{T^2}{T^2 - 9.71 \times T + 60590} \right)^2 \ \text{(J/Kmol)} \qquad (4.5.2)$$

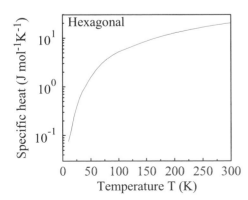

FIG. 4.5.9. Temperature dependence of the specific heat of hexagonal BN [Gorbunov et al. (1988); see also Sichel et al. (1976)].

At $1300 < T < 2200$ K, the specific heat C_p of hexagonal BN can be approximated as [Solozhenko (1994)]

$$C_p = 52.48 - 9.42 \times 10^{-4} \times T - 64877 \times T^{-2} \ (\text{J/Kmol}) \quad (4.5.3)$$

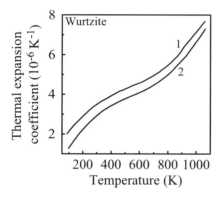

FIG. 4.5.10. Linear thermal expansion coefficient of wurtzite BN parallel (curve 1) and perpendicular (curve 2) to c axis [Kolupayeva et al. (1986)].

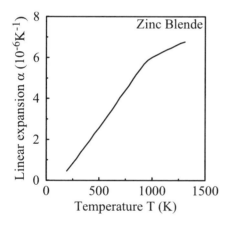

FIG. 4.5.11. Linear thermal expansion coefficient of zinc blende BN [Slack and Bartram (1975)].

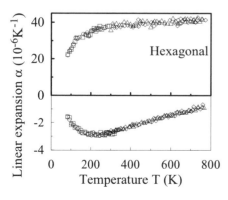

FIG. 4.5.12. Linear thermal expansion coefficients of hexagonal BN. Top curve, in a direction parallel to the c axis; bottom curve, in a direction perpendicular to the c axis [Yates et al. (1975); see also Belenkii et al. (1985)].

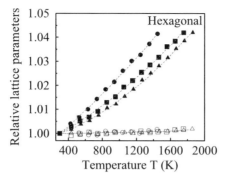

FIG. 4.5.13. Thermal expansion of hexagonal BN at different pressure: circles, 1.6 GPa; squares, 5.0 GPa; triangles, 7.1 GPa. Solid and open symbols are used for the directions parallel and perpendicular to *c* axis, respectively [Solozhenko and Peun (1997)]. Reprinted from *J. Phys. Chem. Solids*, **58**, Solozhenko V.L., and Peun, T., "Compression and thermal expansion of hexagonal graphite-like boron nitride up to 7 GPa and 1800 K," pp. 1321–1323, copyright © 1997 with permission from Elsevier Science.

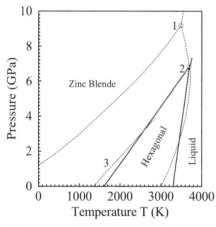

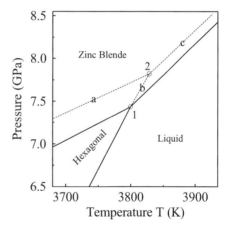

FIG. 4.5.14. Phase diagrams for BN. 1, Bundy–Wentorf's diagram; 2, Equilibrium diagram; 3, h-BN ⇔ c-BN boundary line [Solozhenko (1994)].

FIG. 4.5.15. Equilibrium phase diagram of BN. 1, hexagonal-zinc blende BN liquid triple point; 2, hexagonal-wurtzite BN liquid metastable triple point; a, line of hexagonal-wurtzite BN metastable equilibrium; b, metastable beam of hexagonal BN melting curve; c, line of wurtzite BN metastable melting [Solozhenko et al. (1998)].

4.6. MECHANICAL PROPERTIES, ELASTIC CONSTANTS, LATTICE VIBRATIONS, OTHER PROPERTIES

Wurtzite BN

Density: 3.48 g cm^{-3}

Surface microhardness
(using Knoop's pyramid test)
[Gielisse (1997)]:   3400 kg mm^{-2}

Elastic constants at 300 K
[Shimada et al. (1998); see also Pesin (1980)]:

C_{11}	982 GPa
C_{12}	134 GPa
C_{13}	74 GPa
C_{33}	1077 GPa
C_{44}	388 GPa

Bulk modulus B_s (compressibility^{-1}):

$$B_s = \frac{C_{33}(C_{11} + C_{12}) - 2(C_{13})^2}{C_{11} + C_{12} - 4C_{13} + 2C_{33}}, \qquad B_s = 400 \text{ GPa}$$

Acoustic Wave Speeds

Wave Propagation Direction	Wave Character	Expression for Wave Speed	Wave Speed (in units of 10^5 cm/s)
[001]	V_L (longitudinal)	$(C_{33}/\rho)^{1/2}$	17.6
	V_T (transverse)	$(C_{44}/\rho)^{1/2}$	10.5
[100]	V_L (longitudinal)	$(C_{11}/\rho)^{1/2}$	16.8
	V_T (transverse, polarization along [001])	$(C_{44}/\rho)^{1/2}$	10.5
	V_T (transverse, polarization along [010])	$[(C_{11} - C_{12})/2\rho]^{1/2}$	11.0

For definitions of the crystallographic directions see Appendix 3.

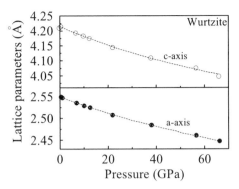

FIG. 4.6.1. Pressure dependence of lattice constants of wurtzite BN [Solozhenko et al. (1998)].

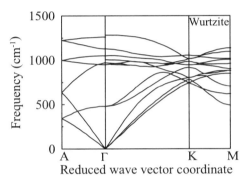

FIG. 4.6.2. Calculated dispersion curves for acoustic and optical branch phonons for wurtzite BN [Karch and Bechstedt (1997)].

Main Phonon Frequencies (in units of cm^{-1}) [Kim et al. (1996), Karch and Bechstedt (1997), Shimada et al. (1998)]

$A_1 - LO$	1258
$A_1 - TO$	1006–1053
$E_1 - LO$	1281
$E_1 - TO$	1053–1085
E_2 (low)	476
E_2 (high)	989

Piezoelectric Constants
[Shimada et al. (1998)]

e_{31}	0.27 C m^{-2}
e_{33}	-0.85 C m^{-2}

Zinc Blende BN

Density: 3.450 g cm^{-3}

Hardness: 9.5 (on the Mohs scale)

Surface microhardness (using Knoop's pyramid test) [Gielisse (1997)]: 4500 kg mm^{-2}

Elastic constants at 300 K [Grimsditch and Zouboulis (1994); see also Shimada et al. (1998)]

C_{11}	820 GPa
C_{12}	190 GPa
C_{44}	480 GPa

Bulk modulus B_s (compressibility^{-1}):

$$B_s = \frac{C_{11} + 2C_{12}}{3} \qquad B_s = 400 \text{ GPa}$$

Anisotropy factor:

$$A = \frac{C_{11} - C_{12}}{2C_{44}} \qquad A = 0.66$$

Shear modulus:

$$C' = (C_{11} - C_{12})/2 \qquad C' = 315 \text{ GPa}$$

[100] Young's modulus Y_0:

$$Y_0 = \frac{(C_{11} + 2C_{12})(C_{11} - C_{12})}{(C_{11} + C_{12})} \qquad Y_0 = 748 \text{ GPa}$$

[100] Poisson ratio σ_0:

$$\sigma_0 = \frac{C_{12}}{C_{11} + C_{12}} \qquad \sigma_0 = 0.19$$

Acoustic Wave Speeds

Wave Propagation Direction	Wave Character	Expression for Wave Speed	Wave Speed (in units of 10^5 cm/s)
[100]	V_L	$(C_{11}/\rho)^{1/2}$	15.4
	V_T	$(C_{44}/\rho)^{1/2}$	11.8
[110]	V_l	$[(C_{11} + C_{12} + 2C_{44})/2\rho)]^{1/2}$	16.9
	$V_{t//}$	$V_{t//} = V_T = (C_{44}/\rho)^{1/2}$	11.8
	$V_{t\perp}$	$[(C_{11} - C_{12})/2\rho)]^{1/2}$	9.6
[111]	V_l'	$[(C_{11} + 2C_{12} + 4C_{44})/3\rho)]^{1/2}$	17.4
	V_t'	$[(C_{11} - C_{12} + C_{44})/3\rho)]^{1/2}$	10.4

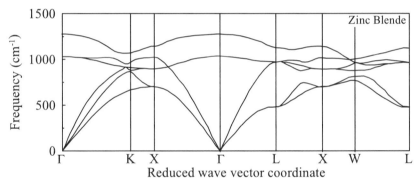

FIG. 4.6.3. Calculated dispersion curves for acoustic and optical branch phonons for zinc blende BN [Karch and Bechstedt (1997)].

Phonon Frequencies (in units of cm^{-1}) [Kim et al. (1996), Karch and Bechstedt (1997), Shimada et al. (1998)]

LO (Γ)	1285
TO (Γ)	1000–1082

Piezoelectric constant
[Shimada et al. (1998)]

e_{14}	-0.64 C m^{-2}

Hexagonal BN

Density: 2.2 g cm^{-3}

Hardness: 1.5 (on the Mohs scale)

Elastic constants at 300 K [Green et al. (1976), Duclaux et al. (1992)]:

C_{11} 750 GPa

C_{12} 150 GPa

C_{33} 32 ± 3 GPa

C_{44} 3 GPa

Bulk modulus B_s (compressibility^{-1})
[Solozhenko and Peun (1997)]: $B_s = 36.5$ GPa

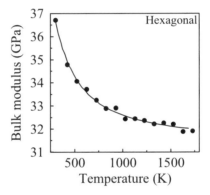

FIG. 4.6.4. Temperature dependence of the bulk modulus of hexagonal BN [Solozhenko and Peun (1997)]. Reprinted from *J. Phys. Chem. Solids* **58**, Solozhenko, V.L., and T. Peun, "Compression and thermal expansion of hexagonal graphite-like boron nitride up to 7 GPa and 1800 K," pp. 1321–1323, copyright © 1997 with permission from Elsevier Science.

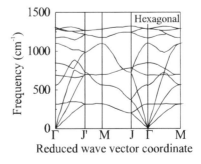

FIG. 4.6.5. Calculated dispersion curves of acoustic and optical branch phonons for hexagonal BN sheet [Miyamoto et al. (1995)].

REFERENCES

Bam, I.S., V.M. Davidenko, V.G. Sidorov, L.I. Fel'dgun, M.D. Shagalov, and Yu.K. Shalabutov, *Sov. Phys. Semicond.* **10**, 3 (1976), 331–332.

Bar-Yam, Y., T. Lei, T.D. Moustakas, D.C. Allen, and M.P. Teter, *Mater. Res. Soc. Symp. Proc.* **242** (1992), 335.

Belenkii, G.L., E.Yu. Salaev, R.A. Suleimanov, N.A. Abdullaev, and V.Ya. Shtein-shraiber, *Solid State Commun.* **53**, 11 (1985), 967–971.

Brozek, T., J. Szmidt, A. Jakubowski, and A. Olszyna, *Diamond Relat. Mater.* **3** (1994), 720–724.

Chrenko, R.M. *Solid State Commun.* **14**, 6 (1974), 511–515.

Christensen, N.E., and I. Gorczyca, *Phys. Rev. B* **50**, 7 (1994), 4397–4415.

Duclaux, L., B. Nysten, J.-P. Issi, and A.W. Moore, *Phys. Rev. B* **46**, 6 (1992), 3362–3367.

Ferhat, M., A. Zaoui, M. Certier, and H. Aourag, *Physica B* **252** (1998), 229–236.

Gielisse, P.J., *Proceedings of the NATO Advanced Research Workshop on Diamond Based Composites and Related Materials*, St. Petersburg, Russia, June 21–22, 1997, pp. 369–373.

Gorbunov, V.E., K.S. Gavrichev, G.A. Totrova, A.V. Bochko, and V.B. Lazarev, *Russian J. Phys. Chem.* **62**, 1 (1988), 9–12.

Green, J.F., T.K. Bolland, and J.W. Bolland, *J. Chem. Phys.* **64**, 2 (1976), 656–662.

Grimsditch, M., and E.S. Zouboulis, *J. Appl. Phys.* **76**, 2 (1994), 832–834.

Gubanov, V.A., E.A. Pentaleri, C.Y. Fong, and B.M. Klein, *Phys. Rev. B* **56**, 20 (1997), 13077–13086.

Hoffman, D.M., G.L. Doll, and P.C. Eklund, *Phys. Rev. B* **30**, 10 (1984), 6051–6056.

Inaba, A., and A. Yoshiasa, *Jpn. J. Appl. Phys.* **36**, Part 1, 9A (1997), 5644–5645.

Karch, K., and F. Bechstedt, *Phys. Rev. B* **56**, 12 (1997), 7404–7415.

Kim, K., W.R.L. Lambrecht, and B. Segall, *Phys. Rev. B* **53**, 24 (1996), 16310–16326.

Kolupayeva, Z.I., M.Ya. Fuks, L.I. Gladkih, A.V. Arinkin, S.V. Malikhin, *J. Less-Common Met.* **117** (1986), 259–263.

Litvinov, D., A. Charles, C.A. Taylor II, and R. Clarke, *Diamond Relat. Mater.* **7** (1998), 360–364.

Lopatin, V.V. *Properties of Group III Nitrides* (edited by Edgar, J.H., EMIS Datareview Series No. 11, INSPEC 1994.

Lu, M., A. Bousetta, K. Waters, and J.A. Shulz, *Appl. Phys. Lett.* **68**, 5 (1996), 622–624.

Lyusternik, V.E., and V.L. Solozhenko, *Russian J. Phys. Chem.* **66**, 5 (1992), 629–631.

Makedon, I.D., A.V. Petrov, and L.I. Fel'dgun, *Izv. Akad. Nauk SSSR, Neorganich. Mater.* **48**, 4 (1972), 765–766.

Mishima, O., J. Tanaka, S. Yamaoka, and O. Fukunaga, *Science* **238** (1987), 181–183.

Miyamoto, Y., L.M.L. Cohen, and S.G. Louie, *Phys. Rev. B* **52**, 20 (1995), 14971–14975.

Miyata, N., K. Moriki, O. Mishima, M. Fujisawa, and T. Hattori, *Phys. Rev. B* **40**, 17 (1989), 12028–12029.

Novikov, N.V., T.D. Osetinskaya, A.A. Shul'zhenko, A.P. Podoba, A.N. Sokolov, I.A. Petrusha, *Dopov. Akad. Nauk. Ukr. RSR, Ser. A Fiz.-Mat. Tekh. Nauki USSR* (1983), 72–75.

Onodera, A., M. Nakatani, M. Kobayashi, Y. Nisida, and O. Mishima, *Phys. Rev. B* **48**, 4 (1993), 2777–2780.

Pesin, V.A. *Sverktverd. Mater.* **6** (1980), 5.

Rodriguez-Hernandez, P., M. Gonzales-Diaz, and A. Munoz, *Phys. Rev. B* **51**, 11 (1995), 14705–14708.

Shimada, K., T. Sota, and K. Suzuku, *J. Appl. Phys.* **84**, 9 (1998), 4951–4958.

Shipilo, V.B., I.P. Guseva, G.V. Leushkina, L.A. Makovetskaya, and G.P. Popel'nyuk, *Inorg. Mater.* **22**, 3 (1986), 361–364.

Sichel, E.K., R.E. Miller, M.S. Abrahams, and C.J. Buiocchi, *Phys. Rev. B* **13**, 10 (1976), 4607–4611.

Simpson, A., and A.D. Stuckes, *J. Phys. C: Solid State Phys.* **4** (1971), 1710–1717.

Sirota, N.N., and N.A. Kofman, *Sov. Phys. Dokl.* **20**, 12 (1975), 861–862.

Sirota, N.N., and N.A. Kofman, *Sov. Phys. Dokl.* **21**, 12 (1976), 516–517.

Slack, G.A., *J. Phys. Chem. Solids* **34** (1973), 321–335.

Slack, G.A., and S.F. Bartram, *J. Appl. Phys.* **46**, 1 (1975), 89–98.

Solozhenko, V.L., V.E. Yachmenev, V.A. Vil'kovskii, A.N. Sokolov, and A.A. Shul'zhenko, *Russian J. Phys. Chem.* **61**, 10 (1987), 1480–1482.

Solozhenko, V.L., *Properties of Group III Nitrides* (edited by Edgar, J.H., EMIS Datareviews Series, No. 11, INSPEC 1994, pp. 43–70.

Solozhenko, V.L., and T. Peun, *J. Phys. Chem. Solids* **58**, 9 (1997), 1321–1323.

Solozhenko, V.L., D. Hausermann, M. Mezouar, and M. Kunz, *Appl. Phys. Lett.* **72**, 14 (1998), 1691–1693.

Stenzel, O., J. Hahn, M. Roder, A. Enrlich, S. Prause, and F. Richter, *Phys. Stat. Sol. (a)* **158** (1996), 281–287.

Taniguchi, T., J. Tanaka, O. Mishima, T. Ohsawa, and S. Yamaoka, *Appl. Phys. Lett.* **62**, 6 (1993), 576–578.

Taylor II, C.A., and R. Clarke, *Proceedings of the NATO Advanced Research Workshop on Diamond Based Composites and Related Materials*, St. Petersburg, Russia, June 21–22, 1997, pp. 63–113.

Waters, K., National Science Foundation Phase II Proposal, February 15, 1995.

Wentorf, Jr., R.F., *J. Chem. Phys.* **26** (1957), 1956.

Yixi, S., J. Xin, W. Kun, S. Chaoshu, H. Zhengfu, S. Junyan, D. Jie, Z. Sheng, and C. Yuanbin, *Phys. Rev. B* **50**, 24 (1994), 18637–18639.

Xu, Yong-Nian, and W.Y. Ching, *Phys. Rev. B* **44**, 15 (1991), 7787–7798.

Yates, B., M.J. Overy, and O. Pirgon, *Philos. Mag.* **32**, 4 (1975), 847–857.

Zunger, A., A. Katzir, and A. Halperin, *Phys. Rev. B* **13**, 12 (1976), 5560–5573.

Silicon Carbide (SiC)

Yu. Goldberg, M. Levinshtein, and S. Rumyantsev
The Ioffe Institute, St. Petersburg, Russia

More than 170 different polytypes of SiC are known. However, about 95% of all publications deal with three main polytypes: $3C$, $4H$, and $6H$.

5.1. BASIC PARAMETERS AT 300 K

	$3C$	$4H$	$6H$
Crystal structure	Zinc blende	Hexagonal	Hexagonal
Group of symmetry	$T_d^2 - F\bar{4}3m$	$C_{6v}{}^4 P6_3mc$	$C_{6v}{}^4 P6_3mc$
Debye temperature (K)	1200	1300	1200
Density (g/cm^3)	3.21	3.21	3.21
Dielectric constant			
static	9.72	The value of $6H$ dielectric constant is usually used	9.66 ($\perp c$ axis) 10.03 ($\|c$ axis)

(Continued)

Properties of Advanced Semiconductor Materials, Edited by Levinshtein, Rumyantsev, Shur.
ISBN 0-471-35827-4 © 2001 John Wiley & Sons, Inc.

	$3C$	$4H$	$6H$
high frequency	6.52	The value of $6H$ dielectric constant is usually used	6.52 ($\perp c$ axis) 6.70 ($\parallel c$ axis)
Effective electron mass (in units of m_0)			
longitudinal m_l/m_0	0.68	0.29	2.0
transverse m_t/m_0	0.25	0.42	0.42
Effective hole masses (in units of m_0)	0.6	~ 1	~ 1
Lattice constants (Å)	4.3596	$a = 3.0730$ $c = 10.053$	$a = 3.0806$ $c = 15.1173$
Optical phonon energy (meV)	102.8	104.2	104.2

Band structure and carrier concentration

	$3C$	$4H$	$6H$
Energy gap (eV)	2.36	3.23	3.0
Conduction band			
Energy separation between Γ_{15v}-valley and L_{1c}-valleys E_L (eV)	4.6		
Energy separation between Γ_{15v}-valley and Γ_{1c}-valley E_Γ (eV)	6.0	5–6	5–6
Energy separation between Γ_{15v} valley and L valleys E_L (eV)		~ 4	
Valence band			
Energy of spin–orbital splitting E_{so} (eV)	0.01	0.007	0.007
Energy of crystal-field splitting E_{cr} (eV)		0.08	0.05
Effective conduction band density of states (cm^{-3})	1.5×10^{19}	1.7×10^{19}	8.9×10^{19}
Effective valence band density of states (cm^{-3})	1.2×10^{19}	2.5×10^{19}	2.5×10^{19}

(Continued)

	3C	4H	6H
Electrical properties	*3C*	*4H*	*6H*
Breakdown field (V cm^{-1})	$\sim 10^6$	$(3 \div 5) \times 10^6$	$(3 \div 5) \times 10^6$
Mobility (cm^2 V^{-1} s^{-1})			
electrons	≤ 800	≤ 900	≤ 400
holes	≤ 320	≤ 120	≤ 90
Diffusion coefficient (cm^2 s^{-1})			
electrons	≤ 20	≤ 22	≤ 10
holes	≤ 8	≤ 3	≤ 2
Electron thermal velocity (m s^{-1})	2×10^5	1.9×10^5	1.5×10^5
Hole thermal velocity (m s^{-1})	1.5×10^5	1.2×10^5	1.2×10^5
Optical properties			
Infrared refractive index	2.55	2.55 ($\perp c$ axis) 2.59 ($\|c$ axis)	2.55 ($\perp c$ axis) 2.59 ($\|c$ axis)
Radiative recombination coefficient (cm^3 s^{-1})		1.5×10^{-12} (estimate)	
Thermal and mechanical properties			
Bulk modulus (dyn cm^{-2})	25×10^{11}	22×10^{11}	22×10^{11} (theoretical estimation 9.7×10^{11} (exp.)
Melting point (°C)	$\sim$3100 K (at 35 atm)	$\sim$3100 K (at 35 atm)	$\sim$3100 K (at 35 atm)
Specific heat (J g^{-1} °C^{-1})	The value of 6H is usually used for estimations	The value of 6H is usually used for estimations	0.69
Thermal conductivity (W cm^{-1} °C^{-1})	3.6	3.7	4.9
Thermal diffusivity (cm^2 s^{-1})	1.6	1.7	2.2
Thermal expansion, linear (°C^{-1})	$\sim 3.8 \times 10^{-6}$		4.3×10^{-6} ($\perp c$ axis) 4.7×10^{-6} ($\|c$ axis)

5.2. BAND STRUCTURE AND CARRIER CONCENTRATION

Comprehensive calculations of the band structure for 3C-, 4H-, and 6H-SiC and related references can be found in Persson and Lindefelt (1997).

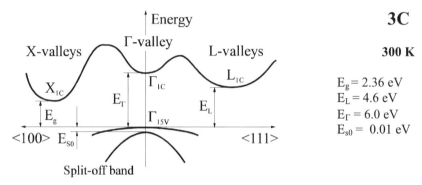

FIG. 5.2.1. Band structure of 3C-SiC. Important minima of the conduction band and maxima of the valence band.

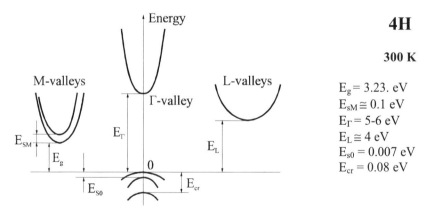

FIG. 5.2.2. Band structure of 4H-SiC. Important minima of the conduction band and maxima of the valence band.

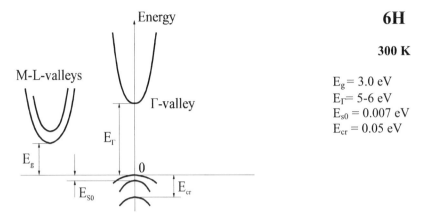

FIG. 5.2.3. Band structure of 6*H*-SiC. Important minima of the conduction band and maxima of the valence band.

5.2.1. Temperature Dependences

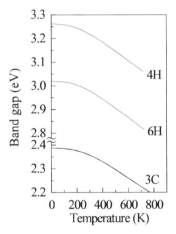

FIG. 5.2.4. The temperature dependences of energy gap for 3*C*-, 4*H*-, and 6*H*-SiC polytypes [Choyke (1969)].

Temperature Dependences of Energy Gap

3C

$$E_g = E_g(0) - 6.0 \times 10^{-4} \times \frac{T^2}{T + 1200} \ (\text{eV}) \qquad (5.2.1)$$

4H

$$E_g = E_g(0) - 6.5 \times 10^{-4} \times \frac{T^2}{T + 1300} \text{ (eV)} \qquad (5.2.2)$$

6H

$$E_g = E_g(0) - 6.5 \times 10^{-4} \times \frac{T^2}{T + 1200} \text{ (eV)} \qquad (5.2.3)$$

where T is temperature in degrees K.

Effective Density of States in the Conduction Band N_c

3C

$$N_c \cong 4.82 \times 10^{15} \cdot M \cdot \left(\frac{m_c}{m_0}\right)^{3/2} \cdot T^{3/2} \text{ (cm}^{-3})$$

$$\cong 4.82 \times 10^{15} \cdot \left(\frac{m_{cd}}{m_0}\right)^{3/2} \cdot T^{3/2} \text{ (cm}^{-3})$$

or

$$N_c \cong 3 \times 10^{15} \times T^{3/2} \text{ (cm}^{-3}) \qquad (5.2.4)$$

$M = 3$ is the number of equivalent valleys in the conduction band.
$m_c = 0.35m_0$ is the effective mass of the density of states in one valley of conduction band.
$m_{cd} = 0.72$ is the effective mass of density of states.

4H

$$N_c \cong 4.82 \times 10^{15} \cdot M \cdot \left(\frac{m_c}{m_0}\right)^{3/2} \cdot T^{3/2} \text{ (cm}^{-3})$$

$$\cong 4.82 \times 10^{15} \cdot \left(\frac{m_{cd}}{m_0}\right)^{3/2} \cdot T^{3/2} \text{ (cm}^{-3})$$

or

$$N_c \cong 3.25 \times 10^{15} \times T^{3/2} \ (\text{cm}^{-3}) \tag{5.2.5}$$

$M = 3$ is the number of equivalent valleys in the conduction band.
$m_c = 0.37m_0$ is the effective mass of the density of states in one valley of the conduction band.
$m_{cd} = 0.77$ is the effective mass of density of states.

6H

$$N_c \cong 4.82 \times 10^{15} \cdot M \cdot \left(\frac{m_c}{m_0}\right)^{3/2} \cdot T^{3/2} \ (\text{cm}^{-3})$$

$$\cong 4.82 \times 10^{15} \cdot \left(\frac{m_{cd}}{m_0}\right)^{3/2} \cdot T^{3/2} \ (\text{cm}^{-3})$$

or

$$N_c \cong 1.73 \times 10^{16} \times T^{3/2} \ (\text{cm}^{-3}) \tag{5.2.6}$$

$M = 6$ is the number of equivalent valleys in the conduction band.
$m_c = 0.71m_0$ is the effective mass of the density of states in one valley of the conduction band.
$m_{cd} = 2.34$ is the effective mass of density of states.

Effective Density of States in the Valence Band, N_v

3C

$$N_v = 2.23 \times 10^{15} \times T^{3/2} \ (\text{cm}^{-3}) \tag{5.2.7}$$

4H

$$N_v \cong 4.8 \times 10^{15} \times T^{3/2} \ (\text{cm}^{-3}) \tag{5.2.8}$$

6H

$$N_v \cong 4.8 \times 10^{15} \times T^{3/2} \ (\text{cm}^{-3}) \tag{5.2.9}$$

Intrinsic carrier concentration:

$$n_i = (N_c \cdot N_v)^{1/2} \exp\left(-\frac{E_g}{2k_B T}\right) \qquad (5.2.10)$$

[see also Ruff et al. (1994), Casady and Johnson (1996)].

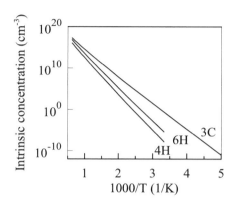

FIG. 5.2.5. The temperature dependences of the intrinsic carrier concentration.

5.2.2. Dependence on Hydrostatic Pressure

For details see Park et al. (1994).

3C

$$E_g = E_g(0) - 0.34 \times 10^{-3} P \text{ (eV)}$$
$$E_L = E_L(0) + 3.92 \times 10^{-3} P \text{ (eV)} \qquad (5.2.11)$$
$$E_\Gamma = E_\Gamma(0) + 5.11 \times 10^{-3} P \text{ (eV)}$$

4H

$$E_g = E_g(0) + 0.08 \times 10^{-3} P \text{ (eV)}$$
$$E_\Gamma = E_\Gamma(0) + 3.7 \times 10^{-3} P \text{ (eV)} \qquad (5.2.12)$$

6H

$$E_g = E_g(0) - 0.03 \times 10^{-3} P \text{ (eV)}$$
$$E_\Gamma = E_\Gamma(0) + 4.03 \times 10^{-3} P \text{ (eV)} \qquad (5.2.13)$$

where P is pressure in kbar.

5.2.3. Energy Gap Narrowing at High Doping Levels
[Lindefelt (1998)]

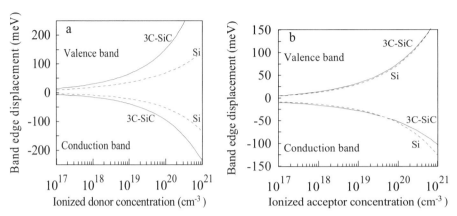

FIG. 5.2.6. Conduction and valence band displacements for $3C$-SiC versus ionized shallow impurity. (a) For n-type material, (b) for p-type material. For comparison, the band-edge displacements for Si are shown.

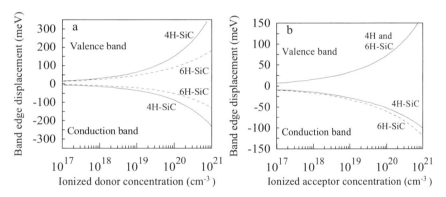

FIG. 5.2.7. Conduction and valence band displacements for $4H$- and $6H$-SiC versus ionized shallow impurity. (a) For n-type material, (b) for p-type material.

The band-edge displacements for n-type material can be expressed with the following equations:

$$\Delta E_c = A_{nc}\left(\frac{N_D^+}{10^{18}}\right)^{1/3} + B_{nc}\left(\frac{N_D^+}{10^{18}}\right)^{1/2} \text{ (eV)}$$

$$\Delta E_v = A_{nv}\left(\frac{N_D^+}{10^{18}}\right)^{1/3} + B_{nv}\left(\frac{N_D^+}{10^{18}}\right)^{1/2} \text{ (eV)}$$

$$(5.2.14)$$

n-type	A_{nc}	B_{nc}	A_{nv}	B_{nv}
Si	-9.74×10^{-3}	-1.39×10^{-3}	1.27×10^{-2}	1.40×10^{-3}
3C-SiC	-1.48×10^{-2}	-3.06×10^{-3}	1.75×10^{-2}	6.85×10^{-3}
4H-SiC	-1.50×10^{-2}	-2.93×10^{-3}	1.90×10^{-2}	8.74×10^{-3}
6H-SiC	-1.12×10^{-2}	-1.01×10^{-3}	2.11×10^{-2}	1.73×10^{-3}

The band-edge displacements for p-type material can be expressed with the following equations:

$$\Delta E_c = A_{pc}\left(\frac{N_A^-}{10^{18}}\right)^{1/4} + B_{pc}\left(\frac{N_A^-}{10^{18}}\right)^{1/2} \ (\text{eV})$$

$$\Delta E_v = A_{pv}\left(\frac{N_A^-}{10^{18}}\right)^{1/3} + B_{pv}\left(\frac{N_A^-}{10^{18}}\right)^{1/2} \ (\text{eV}) \qquad (5.2.15)$$

p-type	A_{pc}	B_{pc}	A_{pv}	B_{pv}
Si	-1.14×10^{-2}	-2.05×10^{-3}	1.11×10^{-2}	2.06×10^{-3}
3C-SiC	-1.50×10^{-2}	-6.41×10^{-4}	1.30×10^{-2}	1.43×10^{-3}
4H-SiC	-1.57×10^{-2}	-3.87×10^{-4}	1.30×10^{-2}	1.15×10^{-3}
6H-SiC	-1.74×10^{-2}	-6.64×10^{-4}	1.30×10^{-2}	1.14×10^{-3}

5.2.4. Effective Masses

For details see Son et al. (1994) and Son et al. (1995).

Electrons

3C The surfaces of equal energy are ellipsoids:

longitudinal: $m_l = 0.68 m_0$
transverse: $m_t = 0.25 m_0$

Effective mass of the density of states
in one valley of conduction band: $m_c = 0.35 m_0$
Effective mass of density of states: $m_{cd} = 0.72 m_0$

There are three equivalent valleys in the conduction band.

Effective mass of conductivity: $m_{cc} = 0.32$

4H The surfaces of equal energy are ellipsoids:

longitudinal: $m_l = 0.29m_0$
transverse: $m_t = 0.42m_0$

Effective mass of the density of states
in one valley of conduction band: $m_c = 0.37m_0$
Effective mass of density of states: $m_{cd} = 0.77m_0$

There are three equivalent valleys in the conduction band.

Effective mass of conductivity: $m_{cc} = 0.36$

6H The surfaces of equal energy are ellipsoids:

longitudinal: $m_l = 2.0m_0$
transverse: $m_t = 0.42m_0$

Effective mass of the density of states
in one valley of conduction band: $m_c = 0.71m_0$
Effective mass of density of states: $m_{cd} = 2.34m_0$

There are six equivalent valleys in the conduction band.

Effective mass of conductivity: $m_{cc} = 0.57m_0$

Holes

3C   Effective mass of density of states: $m_v = 0.6m_0$
4H   Effective mass of density of states: $m_v = 1.0m_0$
6H   Effective mass of density of states: $m_v = 1.0m_0$

5.2.5. Donors and Acceptors

Ionization energies of shallow donors (eV)

	3C	4H	6H
N	0.06–0.1 (0.0536 at 6 K)	0.059–0.102 [Choyke and Pensl (1997)]	0.085–0.125 [Choyke and Pensl (1997)]
Ti		0.13 0.17	
Cr		0.15–0.18 0.74	
P			0.085 0.135

See also a review: Lebedev (1999).

Ionization energies of shallow acceptors (eV)

	3C	4H	6H
Al	0.26	0.19	0.239–0.249 Ikeda et al. (1980)
B			0.27 0.31–0.38 Evwaraye et al. (1997)
Ga	0.344 Kuwabara and Yamada (1975)	0.267 Ikeda et al. (1980)	0.317–0.333 Ikeda et al. (1980)

See also a review: Lebedev (1999).

3C

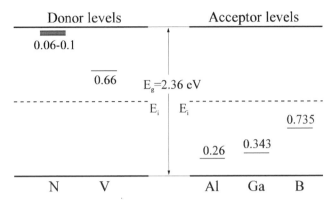

FIG. 5.2.8. The diagram of main impurities in 3C-SiC [Dean et al. (1977), Dombrowski et al. (1994), Ikeda et al. (1980), Kuwabara et al. (1976), Kuwabara and Yamada (1975), Lebedev (1999)].

4H

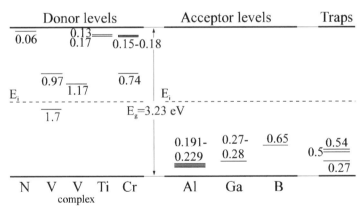

FIG. 5.2.9. The diagram of main impurities in 4H-SiC [Achtziger and Witthuhn (1997), Dalibor et al. (1997), Dombrowski et al. (1994), Evwaraye et al. (1996), Ikeda et al. (1980), Kuznetsov and Zubrilov (1995), Lebedev and Poletaev (1996), Lebedev (1999)].

6H

In all main polytypes of SiC, some atoms have been observed in association both with cubic (c) and with hexagonal (h) lattice sites.

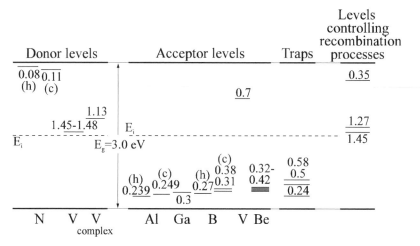

FIG. 5.2.10. The diagram of main impurities in $6H$-SiC [Anikin et al. (1991), Evwaraye et al. (1994), Evwaraye et al. (1996), Evwaraye et al. (1997), Hobgood et al. (1995), Kuznetsov et al. (1994), Kuznetsov and Edmond (1997), Lebedev and Davydov (1997), Mitchel et al. (1997), Troffer et al. (1997), Lebedev (1999)].

5.3. ELECTRICAL PROPERTIES

5.3.1. Mobility and Hall Effect

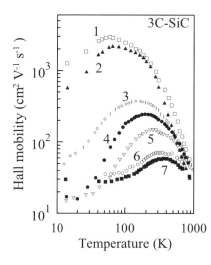

FIG. 5.3.1. Electron Hall mobility versus temperature for different doping levels and different levels of compensation. 3C-SiC. 1, $n_0 \cong 10^{16}$ cm^{-3} at 300 K; 2, $n_0 \cong 5 \times 10^{16}$ cm^{-3} at 300 K; 3, $N_d \cong 1.8 \times 10^{18}$ cm^{-3}, $N_a \cong 1.1 \times 10^{18}$ cm^{-3}; 4, $N_d \cong 3.9 \times 10^{18}$ cm^{-3}, $N_a \cong 2.7 \times 10^{18}$ cm^{-3}; 5, $N_d \cong 6.5 \times 10^{18}$ cm^{-3}, $N_a \cong 3.0 \times 10^{18}$ cm^{-3}; 6, $N_d \cong 8 \times 10^{18}$ cm^{-3}; 7, $N_d \cong 10^{19}$ cm^{-3} [Yamanaka et al. (1987b), Shinohara et al. (1988)].

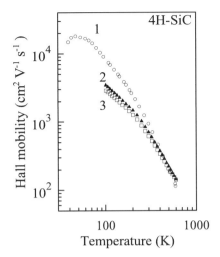

FIG. 5.3.2. Electron Hall mobility versus temperature for different doping levels and different crystallographic orientations for $4H$-SiC. 1, high quality, unintentionally doped; 2, $N_d \cong 1.2 \times 10^{17}$ cm^{-3} electric field $F \| c$; 3, $N_d \cong 1.2 \times 10^{17}$ cm^{-3} electric field $F \perp c$ [Choyke and Pensl (1997), Shaffer et al. (1994)].

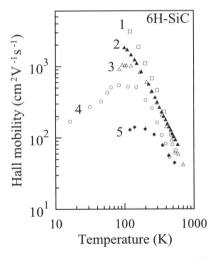

FIG. 5.3.3. Electron Hall mobility versus temperature for different doping levels for $6H$-SiC. 1, $N_d \cong 10^{16}$ cm^{-3}; 2, $N_d \cong 1.5 \times 10^{17}$ cm^{-3}; 3, $N_d \cong 3 \times 10^{17}$ cm^{-3}; 4, $N_d \cong 5 \times 10^{17}$ cm^{-3}; 5, $N_d \cong 1.2 \times 10^{19}$ cm^{-3} [Mickevicius and Zhao (1998), Chen et al. (1996), Barrett and Campbell (1967)].

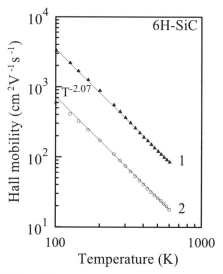

FIG. 5.3.4. Electron Hall mobility versus temperature for different crystallographic orientations. $6H$-SiC. 1, $N_d \cong 10^{17}$ cm^{-3}, electric field $F \perp c$; 2, $N_d \cong 10^{17}$ cm^{-3}, electric field $F \| c$ [Shaffer et al. (1994)].

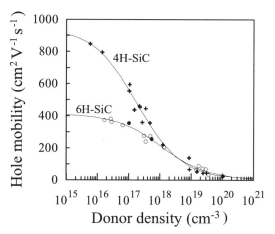

FIG. 5.3.5. Electron Hall mobility versus donor density for $4H$-SiC and $6H$-SiC at $T = 300$ K [Shaffer et al. (1994)].

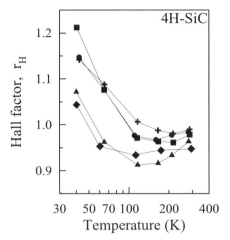

FIG. 5.3.6. The electron Hall factor versus temperature for five 4H-SiC samples with donor concentration $N_d \cong 10^{15}$–3×10^{16} cm^{-3} [Rutsch et al. (1998)].

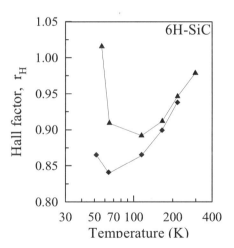

FIG. 5.3.7. The electron Hall factor versus temperature for two 6H-SiC samples $N_d \cong 10^{15}$–10^{16} cm^{-3} [Rutsch et al. (1998)].

See also Chen et al. (1996) and Kinoshita et al. (1997).

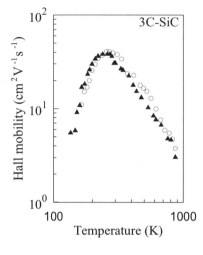

FIG. 5.3.8. Hole Hall mobility versus temperature for two different doping levels for 3C-SiC. Circles, $N_a \cong 5.5 \times 10^{18}$ cm^{-3} and $N_d \cong 3.5 \times 10^{18}$ cm^{-3}; Triangles, $N_a \cong 2 \times 10^{19}$ cm^{-3} and $N_d \cong 5 \times 10^{18}$ cm^{-3} [Yamanaka et al. (1987a)].

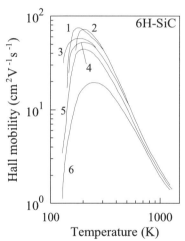

FIG. 5.3.9. Hole Hall mobility versus temperature for different doping levels for 6H-SiC. 1— $N_a \cong 2.3 \times 10^{18}$ cm^{-3}; $N_d \cong 7 \times 10^{17}$ cm^{-3}; 2— $N_a \cong 5.5 \times 10^{18}$ cm^{-3}; $N_d \cong 2.5 \times 10^{18}$ cm^{-3}; 3—$N_a \cong 3.6 \times 10^{18}$ cm^{-3}; $N_d \cong 9 \times 10^{17}$ cm^{-3}; 4—$N_a \cong 1.2 \times 10^{19}$ cm^{-3}; $N_d \cong 3 \times 10^{18}$ cm^{-3}; 5—$N_a \cong 8.5 \times 10^{19}$ cm^{-3}; $N_d \cong 1.4 \times 10^{19}$ cm^{-3}; 6—$N_a \cong 3.6 \times 10^{20}$ cm^{-3}; $N_d \cong 1.2 \times 10^{18}$ cm^{-3} [Van Daal et al. (1963)].

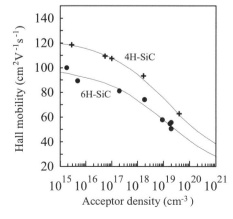

FIG. 5.3.10. Hole Hall mobility versus acceptor density for 4H-SiC and 6H-SiC at $T = 300$ K [Shaffer et al. (1994)].

5.3.2. **Transport Properties in High Electric Field**

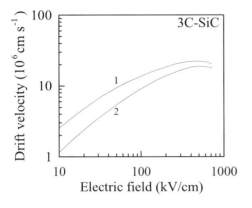

FIG. **5.3.11.** Calculated electron steady-state drift velocity in 3C-SiC as a function of electric field. 1, 300 K; 2, 600 K. Total concentration of ionized dopants is 1.5×10^{18} cm^{-3} [Mickevicius and Zhao (1998)].

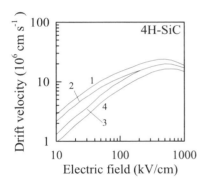

FIG. **5.3.12.** Calculated electron steady state drift velocity of 4H-SiC as a function of electric field. 1 and 2, 300 K; 3 and 4, 600 K. 1 and 3, $E\|c$; 2 and 4, $E\perp c$. Total concentration of ionized dopants is 1.5×10^{18} cm^{-3} [Mickevicius and Zhao (1998)].

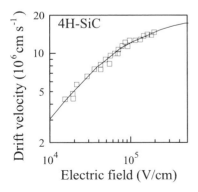

FIG. **5.3.13.** Experimental electron steady state drift velocity of 4H-SiC as function of electric field. 300 K. Mobility in the low electric field is 400 cm^2/V · s. Electron concentration is 1.4×10^{17} cm^{-3} [Khan and Cooper (1998)].

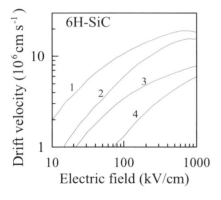

FIG. 5.3.14. Calculated electron steady-state drift velocity of 6*H*-SiC as a function of electric field. 1 and 2, 300 K; 3 and 4, 600 K. 1 and 3, $F \perp c$. 2 and 4, $F \| c$. The total concentration of ionized dopants is 1.5×10^{18} cm^{-3} [Mickevicius and Zhao (1998)].

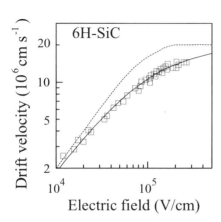

FIG. 5.3.15. Experimental electron steady-state drift velocity of 6*H*-SiC as a function of electric field at 300 K. Mobility in low electric field is 200 cm^2/V · s. Electron concentration is 1.2×10^{17} cm^{-3} [Khan and Cooper (1998)]. Dotted line represents the results by von Muench and Pettenpaul (1977).

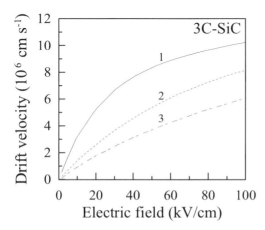

FIG. 5.3.16. Calculated hole steady-state drift velocity of 3*C*-SiC as a function of electric field applied along (100) direction (relatively low fields). 300 K. Impurity concentration (cm^{-3}): 1, 10^{17}; 2, 10^{18}; 3, 7×10^{18} [Bellotti et al. (1999)].

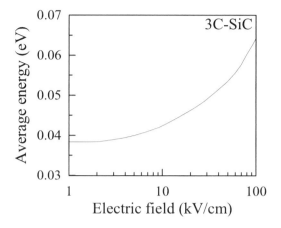

FIG. 5.3.17. Calculated average hole energy of 3C-SiC as a function of electric field applied along (100) direction (relatively low fields) at 300 K [Bellotti et al. (1999)].

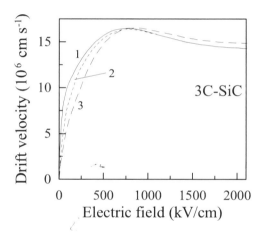

FIG. 5.3.18. Calculated hole steady-state drift velocity of 3C-SiC as a function of electric field applied along (100) direction (high fields) at 300 K. Impurity concentration (cm^{-3}): 1, 10^{17}; 2, 10^{18}; 3, 7 × 10^{18} [Bellotti et al. (1999)].

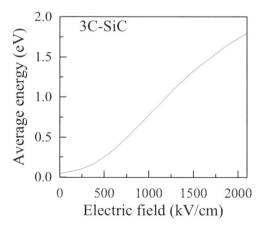

FIG. 5.3.19. Calculated average hole energy of 3C-SiC as a function of electric field applied along (100) direction (high fields) [Bellotti et al. (1999)].

5.3.3. Impact Ionization

3C

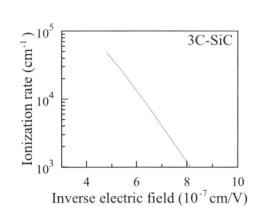

FIG. 5.3.20. Calculated hole ionization rates as a function of inverse electric field in 3*C*-SiC at 300 K [Bellotti et al. (1999)].

4H

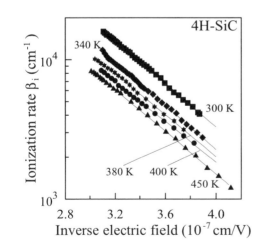

FIG. 5.3.21. Measured hole ionization rates as a function of inverse electric field in 4*H*-SiC at different temperatures [Raghunathan and Baliga (1999)]. Reprinted from *Solid State Electronics* **43**, Raghunathan, R., and Baliga, B., "Temperature dependence of hole impact ionization coefficients in 4*H* and 6*H*-SiC," pp. 199–211, copyright © (1999), with permission from Elsevier Science.

$$\beta_i = \beta_0 \times \exp[-F_{po}/F]$$

$$\beta_0 = 6.3 \times 10^6 - 1.07 \times 10^4 T \ (\text{cm}^{-1})$$

$$F_{po} = 1.8 \times 10^7 \ (\text{V/cm})$$

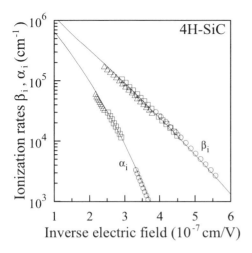

FIG. 5.3.22. Ionization rates for electrons and holes in 4*H*-SiC as a function of inverse electric field at 300 K [Konstantinov et al. (1997)].

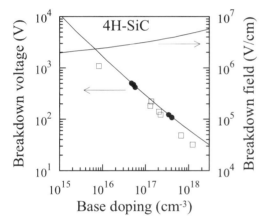

FIG. 5.3.23. Dependences of the breakdown voltage and breakdown field on base doping level for 4*H*-SiC abrupt p^+–n junctions at 300 K [Konstantinov et al. (1997)].

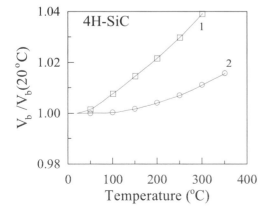

FIG. 5.3.24. Normalized temperature dependences of the breakdown voltage of 4*H*-SiC for uniform breakdown of abrupt p^+–n junction. Breakdown voltage (V): 1, 452; 2, 107 [Konstaninov et al. (1998)].

6H

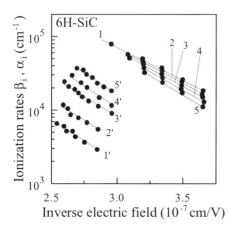

FIG. 5.3.25. Electron (lines $1'$–$5'$) and hole (lines 1–5) ionization rates as a function of inverse electric field in 6H-SiC at different temperatures. T (K): 1 and $1'$, 294; 2 and $2'$, 373; 3 and $3'$, 473; 4 and $4'$, 573; 5 and $5'$, 673 [Konstantinov (1989)].

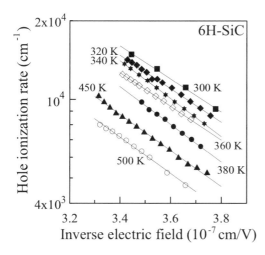

FIG. 5.3.26. Hole ionization rates as a function of inverse electric field in 6H-SiC at different temperatures [Raghunathan and Baliga (1999)]. Reprinted from *Solid State Electronics* **43**, Raghunathan, R., and Baliga, B., "Temperature dependence of hole impact ionization coefficients in 4H and 6H-SiC," pp. 199–211, copyright © 1999, with permission from Elsevier Science.

Electron ionization rates (300 K):

$$\alpha_i = \alpha_0 \times \exp(-F_i/F),$$

where $\alpha_0 = 4.57 \times 10^8$ cm^{-1}, $\quad F_i = 5.24 \times 10^7$ V/cm

Hole ionization rates (300 K):

$$\beta_i = \beta_0 \times \exp(-F_i/F),$$

$$\text{where } \beta_0 = 5.13 \times 10^6 \text{ cm}^{-1}, \quad F_i = 1.57 \times 10^7 \text{ V/cm}$$

[Kyuregyan and Yurkov (1989)].

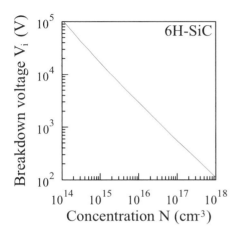

FIG. 5.3.27. Dependence of the breakdown voltage for 6*H*-SiC abrupt $p^+–n$ junctions at 300 K [Kyuregyan and Yurkov (1989)].

Temperature Dependence of Breakdown Voltage in 6H-SiC

Temperature coefficient of the hole ionization rates is negative (temperature coefficient of breakdown voltage of 6*H*-SiC is positive) in the material with the small concentration of the traps. On the other hand, in the material with large concentration of the traps, the temperature coefficient of the hole ionization rates is positive.

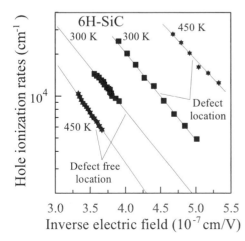

FIG. 5.3.28. Hole ionization rates as a function of inverse electric field in 6*H*-SiC at two temperatures for defective and defective free materials [Raghunathan and Baliga (1998)].

Temperature coefficient of breakdown voltage in n-type of $6H$-SiC is positive for the field direction $F \perp c$ in the material with the small concentration of the traps.

There are two schools of thought regarding the temperature dependence of breakdown voltage in $6H$-SiC for the field direction $F \| c$. The first one explains the negative temperature coefficient by the discontinuity of the electron energy spectrum for motion along c axis [Konstantinov (1989), Konstantinov et al. (1998)]. The second school states that the negative temperature coefficient of breakdown for $F \| c$ is attributed to the traps in the material.

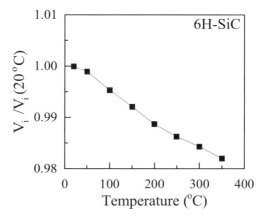

FIG. 5.3.29. Temperature dependence of normalized breakdown voltage for $6H$-SiC. $F \| c$ [Konstantinov et al. (1998)].

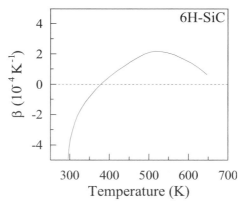

FIG. 5.3.30. Temperature dependence of breakdown volage temperature coefficient $\beta = (d/dT)(\ln V_i)$ for $6H$-SiC. $F \| c$ [Anikin et al. (1988)].

5.3.4. Recombination Parameters

4H

Pure *n*-type material (300 K):
    The longest lifetime of holes: $\tau_p \approx 6 \times 10^{-7} c$
    Diffusion length $L_p = (D_p \times \tau_p)^{1/2}$: $L_p \approx 12$ μm
Pure *p*-type material (300 K):
    The longest lifetime of electrons: $\tau_n \approx 10^{-9} c$
    Diffusion length $L_n = (D_n \times \tau_n)^{1/2}$: $L_n \approx 1.5$ μm

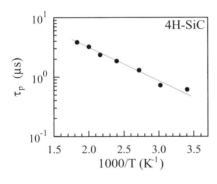

FIG. 5.3.31. The dependence of hole lifetime τ_p on reciprocal temperature measured in the base of a high-voltage *4H*-SiC rectifier diode [Ivanov et al. (1999)].

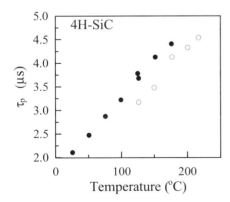

FIG. 5.3.32. The dependence of hole lifetime τ_p on temperature measured using the photoluminescence decay technique [Kordina et al. (1996)].

Surface Recombinaton Velocity

Surface recombination velocity lies in the range from 10^3 to 10^5 cm/s [Galeskas et al. (1997), Grivickas et al. (1997), Neudeck and Fazi (1997), Kimoto et al. (1999)].

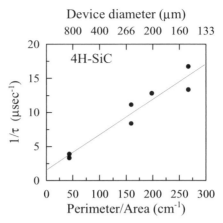

FIG. 5.3.33. The dependence of reciprocal effective hole lifetime $1/\tau$ on perimeter-to-area P/A ratio for $4H$-SiC p–n structure. The slope of the dependence defines the surface recombination velocity according to the equation $1/\tau = 1/\tau_p + s \cdot (A/P)$, where τ_p is the volume hole lifetime [Neudeck and Fazi (1997)].

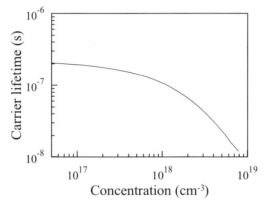

FIG. 5.3.34. The carriers lifetime as a function of carrier density. Solid curve is calculated to fit experimental data. $\tau = 2.6 \times 10^{-7}$ s, $B = 1.5 \times 10^{-12}$ cm^3 s^{-1}, and $C = 7 \times 10^{-31}$ cm^6 s^{-1} [Galeskas et al. (1997)].

Radiative recombination coefficient B
at 300 K [Galeskas et al. (1997)]:

1.5×10^{-12} cm^3 s^{-1} (estimation)
    Auger coefficient at 300 K [Galeskas et al. (1997)]:

C_n	5×10^{-31} cm^6 s^{-1}
C_p	2×10^{-31} cm^6 s^{-1}
$C = C_n + C_p$	7×10^{-31} cm^6 s^{-1}

6H

Pure *n*-type material (300 K):
  The longest lifetime of holes: $\tau_p \approx 4.5 \times 10^{-7} c$
  Diffusion length $L_p = (D_p \times \tau_p)^{1/2}$: $L_p \approx 10$ μm
Pure *p*-type material (300 K):
  The longest lifetime of electrons: $\tau_n \approx 10^{-9} c$
  Diffusion length $L_n = (D_n \times \tau_n)^{1/2}$: $L_n \approx 1.0$ μm

Surface recombinaton velocity lies in the range from 10^4–10^5 cm/s.

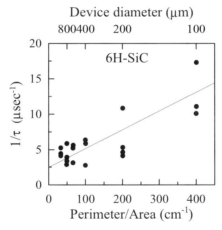

FIG. 5.3.35. The dependence of reciprocal effective hole lifetime $1/\tau$ on perimeter-to-area P/A ratio for 6*H*-SiC *p–n* structure. The slope of the dependence defines the surface recombination velocity according to the equation $1/\tau = 1/\tau_p + s \cdot (A/P)$, where τ_p is the volume hole lifetime [Kimoto et al. (1999)]. Reprinted from Kimoto, T., Miyamoto, N., and Matsunami, H., "Performance limiting surface defects in SiC epitaxial *p–n* junction diodes," *IEEE Trans. Electron Dev.* **46**, 3 (1999), 471–477, copyright © 1999 IEEE.

The values of radiative recombination coefficient B and Auger co-efficients C measured in 4*H*-SiC can be used for estimations.

Radiative recombination coefficient B
at 300 K [Galeskas et al. (1997)]: $\sim 1.5 \times 10^{-12}$ cm^3 s^{-1}
Auger coefficient at 300 K [Galeskas et al. (1997)]:
  C_n $\sim 5 \times 10^{-31}$ cm^6 s^{-1}
  C_p $\sim 2 \times 10^{-31}$ cm^6 s^{-1}
  $C = C_n + C_p$ $\sim 7 \times 10^{-31}$ cm^6 s^{-1}

5.4. OPTICAL PROPERTIES

3*C*

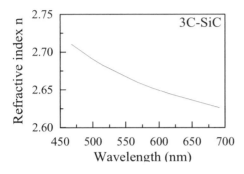

FIG. 5.4.1. Refractive index *n* versus wavelength of 3*C*-SiC 300 K [Shaffer (1971)].

Infrared refractive index (300 K):

$$n_\infty \approx (k_\infty)^{1/2} \approx 2.55$$

For 467 nm $< \lambda <$ 691 nm (300 K):

$$n(\lambda) = 2.55378 + 3.417 \times 10^4 \cdot \lambda^{-2} \quad \text{[Shaffer and Naum (1969)]}$$

where λ is the wavelength in nm.

4*H*

Infrared refractive index (300 K):

$$n_\infty \approx (k_\infty)^{1/2} \approx 2.55 \ (\perp c \text{ axis})$$
$$n_\infty \approx (k_\infty)^{1/2} \approx 2.59 \ (\|c \text{ axis})$$

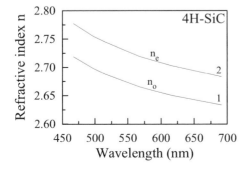

FIG. 5.4.2. Refractive index *n* versus wavelength of 4*H*-SiC for the directions $\perp c$ axis (curve 1) and $\|c$ axis (curve 2) at 300 K [Shaffer (1971)].

$$n_0(\lambda) = 2.5610 + 3.4 \times 10^4 \cdot \lambda^{-2}$$

$$n_e(\lambda) = 2.6041 + 3.75 \times 10^4 \cdot \lambda^{-2}$$

where λ is the wavelength in nm [Shaffer (1971)].

6H

Infrared refractive index (300 K):

$$n_\infty \approx (k_\infty)^{1/2} \approx 2.55 \; (\perp c \text{ axis})$$

$$n_\infty \approx (k_\infty)^{1/2} \approx 2.59 \; (\| c \text{ axis})$$

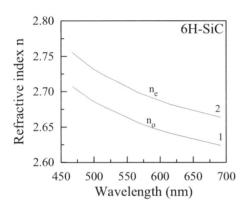

FIG. 5.4.3. Refractive index n versus wavelength of 6H-SiC for the directions $\perp c$ axis (curve 1) and $\| c$ axis (curve 2) at 300 K [Shaffer (1971)].

$$n_0(\lambda) = 2.5531 + 3.34 \times 10^4 \cdot \lambda^{-2}$$

$$n_e(\lambda) = 2.5852 + 3.68 \times 10^4 \cdot \lambda^{-2}$$

where λ is the wavelength in nm [Shaffer (1971)].

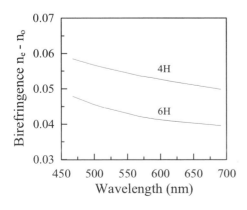

FIG. 5.4.4. Birefringence $(n_e - n_0)$ of 4H- and 6H-SiC versus wavelength [Shaffer (1971)].

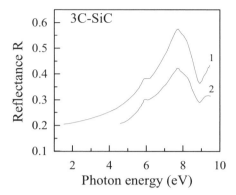

FIG. 5.4.5. Reflectance R as a function of photon energy for 3C-SiC at 300 K. Curve 1, Logothetidis and Petalas (1996); curve 2, Lambrecht et al. (1994).

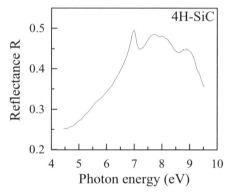

FIG. 5.4.6. Reflectance R as a function of photon energy for 4H-SiC at 300 K [Lambrecht et al. (1993)].

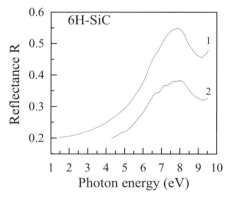

FIG. 5.4.7. Reflectance R as a function of photon energy for 6H-SiC at 300 K. Curve 1, Logothetidis and Petalas (1996); curve 2, Lambrecht et al. (1994).

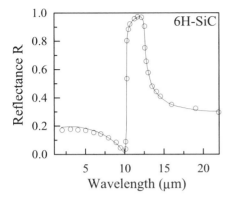

FIG. 5.4.8. Reflectance R as a function of wavelength for $6H$-SiC at 300 K. $F \perp c$ [Spitzer et al. (1959)].

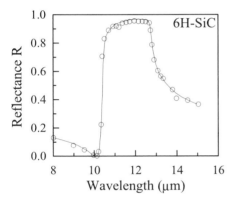

FIG. 5.4.9. Reflectance R as a function of wavelength for $6H$-SiC at 300 K. $F \| c$ [Spitzer et al. (1959)].

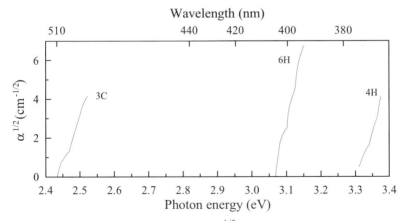

FIG. 5.4.10. The dependences of absorption $\alpha^{1/2}$ versus photon energy for $3C$-, $4H$-, and $6H$-SiC polytypes at 4.2 K. Light-polarized $F \perp c$ [Choyke (1969)].

3C

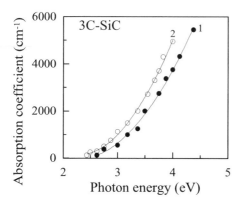

FIG. **5.4.11.** The absorption coefficient versus photon energy for 3*C*-SiC samples at different electron concentrations at 300 K. N_d (cm^{-3}): Curve 1, 5×10^{16}; curve 2, 7×10^{16}. Experimental points are taken from Solangi and Chaudhry (1992). Solid lines: $\alpha \sim (h\nu)^2$.

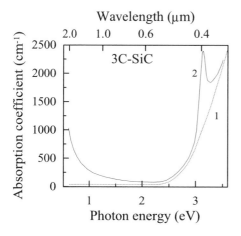

FIG. **5.4.12.** The absorption coefficient versus photon energy for 3*C*-SiC samples at different electron concentrations at 300 K. Curve 1, relatively pure crystal; curve 2, $N_d \cong 10^{19}$ cm^{-3} [Patrick and Choyke (1969)].

4H

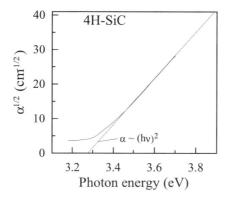

FIG. **5.4.13.** The absorption coefficient versus photon energy for 4*H*-SiC at 300 K. $F \perp c$. Low-doped samples [Sridhara et al. (1998)].

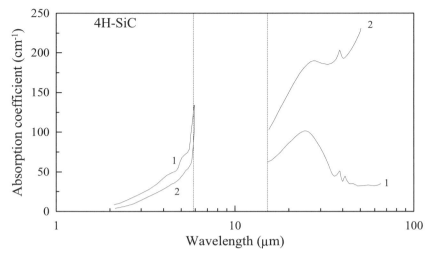

FIG. 5.4.14. The infrared absorption coefficient versus wavelength for 4H-SiC. T (K): Curve 1, 80; curve 2, 300. $F \perp c$. $N_d - N_a \cong 3 \times 10^{17}$ cm^{-3} [Radovanova (1973)].

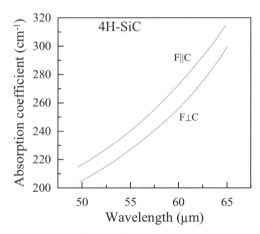

FIG. 5.4.15. Free carrier absorption coefficient versus wavelength of 4H-SiC for $F \perp c$ and $F \| c$ at 300 K. $N_d - N_a \cong 3 \times 10^{17}$ cm^{-3} [Radovanova (1973)].

6H

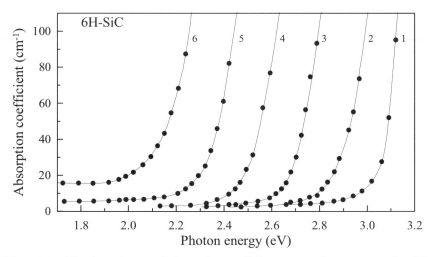

FIG. 5.4.16. The dependences of absorption coefficient versus photon energy for *6H*-SiC at different temperatures. T (°C): Curve 1, 20; curve 2, 300; curve 3, 600; curve 4, 900; curve 5, 1200; curve 6, 1500 [Groth and Kauer (1961)].

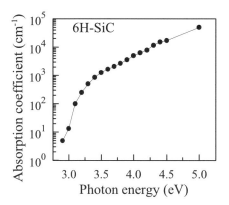

FIG. 5.4.17. The dependence of absorption coefficient versus photon energy. *6H*-SiC at 300 K [Philipp and Taft (1960)].

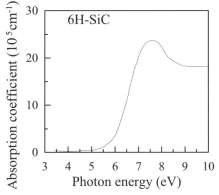

FIG. 5.4.18. The dependence of absorption coefficient versus photon energy. *6H*-SiC at 300 K [Philipp and Taft (1960)].

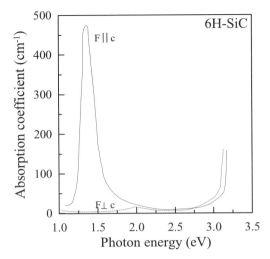

FIG. 5.4.19. The dependences of absorption coefficient versus photon energy of 6*H*-SiC for $F\|c$ and $F\perp c$ at 300 K. $N_d - N_a \cong 3.4 \times 10^{18}$ cm^{-3} [Radovanova (1973)].

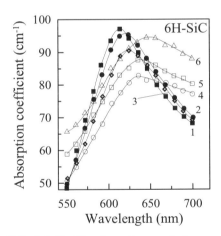

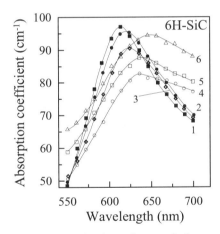

FIG. 5.4.20. The dependences of absorption coefficient versus photon energy of 6*H*-SiC for $F\|c$ at different temperatures. T (K): Curve 1, 78; curve 2, 300; curve 3, 390; curve 4, 550; curve 5, 810. $N_d - N_a \cong 2.0 \times 10^{18}$ cm^{-3} [Dubrovskii et al. (1973)].

FIG. 5.4.21. The dependences of absorption coefficient versus wavelength of 6*H*-SiC for $F\perp c$ at different temperatures. T (K): Curve 1, 80; curve 2, 300; curve 3, 450; curve 4, 640; curve 5, 930; curve 6, 1100. $N_d - N_a \cong 1 \times 10^{19}$ cm^{-3} [Dubrovskii and Radovanova (1968)].

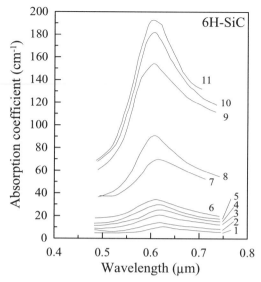

FIG. 5.4.22. The dependences of absorption coefficient versus wavelength of $6H$-SiC for $F \perp c$ at different $N_d - N_a$ values at 300 K. $N_d - N_a$ (cm^{-3}): Curve 1, 1.6×10^{18}; curve 2, 2.7×10^{18}; curve 3, 4.0×10^{18}; curve 4, 4.8×10^{18}; curve 5, 5.8×10^{18}; curve 6, 6.3×10^{18}; curve 7, 1.3×10^{19}; curve 8, 1.6×10^{19}; curve 9, 2.6×10^{19}; curve 10, 3.3×10^{19}; curve 11, 4.0×10^{19} [Radovanova (1973)].

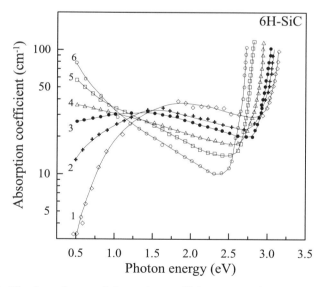

FIG. 5.4.23. The dependences of absorption coefficient versus photon energy of $6H$-SiC doped with B at different temperatures. T (K): Curve 1, 300; curve 2, 400; curve 3, 500; curve 4, 600; curve 5, 700; curve 6, 800 [Tairov and Tsvetkov (1988)].

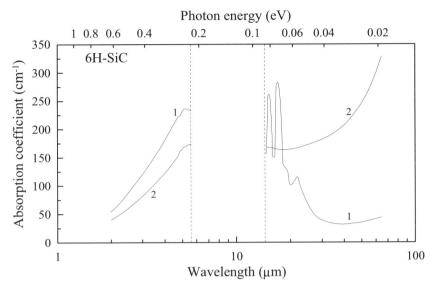

FIG. 5.4.24. The infrared absorption coefficient versus wavelength of $6H$-SiC. T (K): 1, 80; 2, 300 [Dubrovskii and Radovanova (1971)].

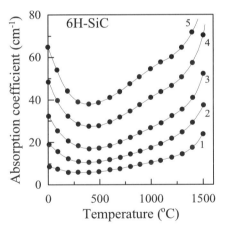

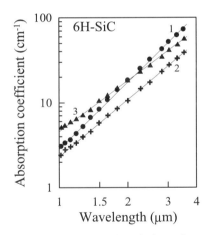

FIG. 5.4.25. The temperature dependences of the absorption coefficient of $6H$-SiC at different wavelength λ. $\lambda(\mu m)$: Curve 1, 1.49; curve 2, 2.0; curve 3, 2.5; curve 4, 3.03; curve 5, 3.52 [Groth and Kauer (1961)].

FIG. 5.4.26. The wavelength dependences of the absorption coefficient α for $6H$-SiC at different temperatures. T (°C): Curve 1, 20, $\alpha \sim \lambda^{2.6}$; curve 2, 400, $\alpha \sim \lambda^{2.2}$; curve 3, 1000, $\alpha \sim \lambda^{1.95}$ [Groth and Kauer (1961)].

5.5. THERMAL PROPERTIES

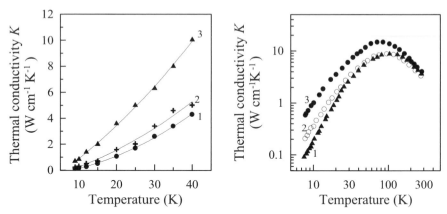

FIG. 5.5.1. The temperature dependences of thermal conductivity at low temperatures. Curve 1, 4*H*; curve 2, 3*C*; curve 3, 6*H* [After Harris (1995a)].

FIG. 5.5.2. The temperature dependences of thermal conductivity. Curve 1, 4*H*; curve 2, 3*C*; curve 3, 6*H* [Morelli et al. (1993)].

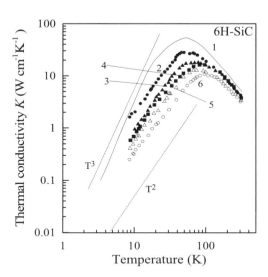

FIG. 5.5.3. Temperature dependences of thermal conductivity for 6*H*-SiC at different electron concentrations. Curve 1, very pure or highly compensated sample. n (cm^{-3}): Curve 2, 3.5×10^{16}; curve 3, 2.5×10^{16}; curve 4, 8×10^{17}; curve 5, 2×10^{17}; curve 6, 3×10^{18} [Morelli et al. (1993)].

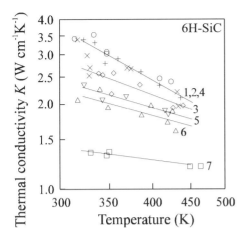

FIG. **5.5.4.** The temperature dependences of thermal conductivity of different 6*H*-SiC samples (normal to the *c* axis). Curves 1, 2, and 3: *n*-type. *n* at 300 K (cm^{-3}): Curve 1, 8×10^{15}; curve 2, 5×10^{16}; curve 3, 1×10^{19}. Curves 4, 5, 6, and 7: *p*-type. *p* at 300 K (cm^{-3}): Curve 4, 2×10^{16}; curve 5, 5×10^{18}; curve 6, 5×10^{19}; curve 7, $\approx 10^{20}$ [Burgemeister et al. (1979)].

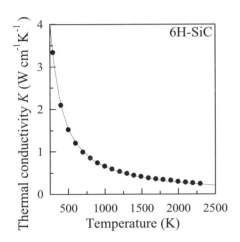

FIG. **5.5.5.** The temperature dependence of thermal conductivity at high temperatures. 6*H*-SiC. Fitted solid line has been calculated according to equation

$$K = \frac{611}{T - 115} \ (\text{W cm}^{-1} \ \text{K}^{-1})$$

where *T* is temperature in degrees K [Nilsson et al. (1997)].

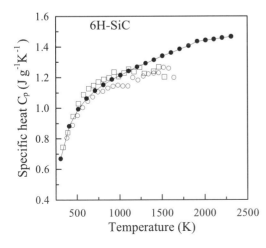

FIG. **5.5.6.** The temperature dependence of specific heat for monocrystalline 6*H*-SiC [Nilsson et al. (1997)].

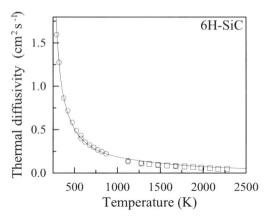

FIG. 5.5.7. The temperature dependence of thermal diffusivity (6H-SiC). Fitted solid line has been calculated according to equation

$$K = \frac{146}{T - 207} \ (\text{cm}^2 \ \text{s}^{-1})$$

where T is temperature in degrees K [Nilsson et al. (1997)].

Thermal Expansion a [Taylor and Jones (1960), Kern et al. (1969)]

3C

Polycrystal: $\alpha \times 10^6$ 2.47

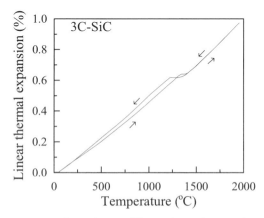

FIG. 5.5.8. The temperature dependence of linear thermal expansion of polycrystal 3C-SiC [Kern et al. (1969)].

Temperature (K)	Single crystal $\alpha \times 10^6$
500	3.8
600	4.3
900	4.8
1500–2100	5.5

6H

Temperature (K)	$\alpha \times 10^6$
100	1.2
700	4.3 ($\perp c$ axis)
700	4.7 ($\parallel c$ axis)

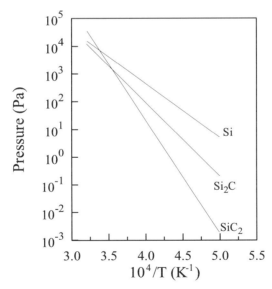

FIG. 5.5.9. Partial pressures of the various species over SiC in SiC-Si system [Tairov and Tsvetkov (1988)].

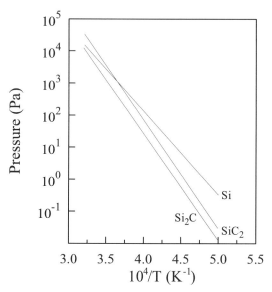

FIG. 5.5.10. Partial pressures of the various species over SiC in SiC-C system [Tairov and Tsvetkov (1988)].

Partial Pressures of the Various Species over SiC (atm)
[Drowart et al. (1958)]

T (K)	Si	SiC	SiC_2	Si_2	Si_2C	Si_2C_2	Si_3	Si_2C_3	Si_3C
2149	2.1×10^{-5}		1.9×10^{-6}	3.8×10^{-8}	1.4×10^{-6}				
2168	2.7×10^{-5}		2.5×10^{-6}	4.8×10^{-8}	1.9×10^{-6}				
2181	3.3×10^{-5}	2.2×10^{-9}	4.2×10^{-6}	6.7×10^{-8}	2.6×10^{-6}				
2196	2.1×10^{-5}		4.4×10^{-6}	1.1×10^{-7}	3.9×10^{-6}	8.5×10^{-9}			1.5×10^{-8}
2230	6.5×10^{-5}		6.5×10^{-6}	1.6×10^{-7}	5.1×10^{-6}	1.6×10^{-8}	3.2×10^{-9}	3.6×10^{-9}	1.8×10^{-8}
2247	8.3×10^{-5}	6.3×10^{-9}	1.1×10^{-5}	2.1×10^{-7}	8.1×10^{-6}				
2316	2.0×10^{-4}	1.9×10^{-8}	3.1×10^{-5}	7.0×10^{-7}	2.2×10^{-5}	7.5×10^{-8}	1.6×10^{-8}	1.7×10^{-8}	8.5×10^{-8}

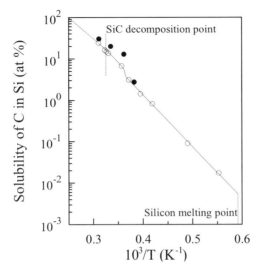

FIG. 5.5.11. Solubility of carbon (C) in silicon (Si) [Marshall (1969)].

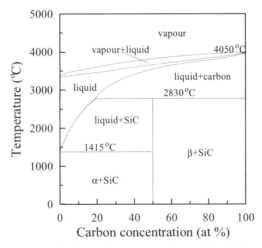

FIG. 5.5.12. Phase diagram in Si–C the system. α is a solid solution of C in Si. β is a solid solution of Si in C [Tairov and Tsvetkov (1988)].

Melting Points T_m

Silicon	1685 K	
Carbon	4100 K	(at $P = 125$ kbar)
SiC	3103 $\pm$ 40 K	(at $P = 35$ atm)

5.6. MECHANICAL PROPERTIES, ELASTIC CONSTANTS, LATTICE VIBRATIONS, OTHER PROPERTIES

Density:	3.21 g cm^{-3} (for details see Harris (1995b)
Hardness:	9.2–9.3 on the Mohs scale
Surface microhardness (using Knoop's pyramid test):	2900–3100 kg mm^{-2} (at 300 K) [Kern et al. (1969), Shaffer (1965)]

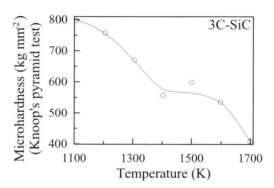

FIG. 5.6.1. Temperature dependence of surface microhardness at elevated temperatures (using Knoop's pyramid test) of polycrystal 3C-SiC [Kern et al. (1969)].

3C-SiC

Elastic constants at 300 K [Gmelins Handbuch (1959)]:

C_{11} 290 GPa
C_{12} 235 GPa
C_{44} 55 GPa

Bulk modulus B_s (compressibility^{-1}):

$$B_s = \frac{C_{11} + 2C_{12}}{3}, \qquad B_s = 250 \text{ GPa}$$

Anisotropy factor

$$A = \frac{C_{11} - C_{12}}{2C_{44}}, \qquad A = 0.5$$

Shear modulus

$$C' = (C_{11} - C_{12})/2, \qquad C' = 27.5 \text{ GPa } [160 \text{ GPa}$$
according to Tairov and Tsvetkov
(1988)]

[100] Young's modulus Y_0:

$$Y_0 = \frac{(C_{11} + 2C_{12})(C_{11} - C_{12})}{C_{11} + C_{12}}, \qquad \begin{array}{l} Y_0 = 80 \text{ GPa} \\ [392\text{--}694 \text{ GPa according to} \\ \text{Harris (1995c)}] \end{array}$$

[100] Poisson ratio σ_0:

$$\sigma_0 = \frac{C_{12}}{C_{11} + C_{12}}, \qquad \sigma_0 = 0.45$$

Acoustic Wave Speeds

Wave Propagation Direction	Wave Character	Expression for Wave Speed	Wave Speed (in units of 10^5 cm/s)
[100]	V_L	$(C_{11}/\rho)^{1/2}$	9.5
	V_T	$(C_{44}/\rho)^{1/2}$	4.1
[110]	V_l	$[(C_{11} + C_{12} + 2C_{44})/2\rho)]^{1/2}$	9.95
	$V_{t\parallel}$	$V_{t\parallel} = V_T = (C_{44}/\rho)^{1/2}$	4.1
	$V_{t\perp}$	$[(C_{11} - C_{12})/2\rho)]^{1/2}$	2.9
[111]	V_l'	$[(C_{11} + 2C_{12} + 4C_{44})/3\rho)]^{1/2}$	10.1
	V_t'	$[(C_{11} - C_{12} + C_{44})/3\rho)]^{1/2}$	3.4

Experimental Results on Acoustic Velocity in 3C-SiC (polycrystal): 1.26×10^6 cm/s [Chinone et al. (1989)]

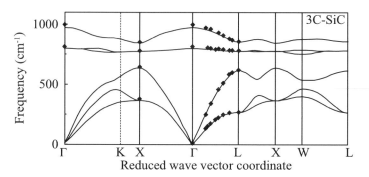

FIG. 5.6.2. Dispersion curves for acoustic and optical branch phonons for 3C-SiC [Karch et al. (1994)].

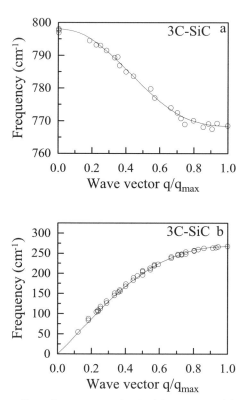

FIG. 5.6.3. Phonon dispersion curves of TO (a) and TA (b) branches of 3C-SiC [Nakashima and Tahara (1989)].

Main Phonon Frequencies (in units of cm^{-1}) [Karch et al. (1994), Nakashima and Tahara (1989), Feldman et al. (1968), Olego and Cardona (1982)]

Γ_{TO}	783–796
Γ_{LO}	829
X_{TO}	755–761
X_{LO}	829
L_{TO}	765–766
L_{LO}	837–838
X_{TA}	366–373
X_{LA}	629–640
L_{TA}	261–266
L_{LA}	610

4*H*- and 6*H*-SiC

The elastic constants in 4*H*- and 6*H*-SiC are the same within experimental uncertainties [Kamitani et al. (1997)].

Elastic constants at 300 K:

C_{11}	501 GPa
C_{12}	111 GPa
C_{13}	52 GPa [see also Frantsevich et al. (1982)]
C_{33}	553 GPa
C_{44}	163 GPa

Bulk modulus B_s (compressibility^{-1}):

$$B_s = \frac{C_{33}(C_{11} + C_{12}) - 2(C_{13})^2}{C_{11} + C_{12} - 4C_{13} + 2C_{33}}, \qquad \begin{array}{l} B_s = 220 \text{ GPa } [97 \text{ GPa} \\ \text{according to Shaffer (1964)}] \end{array}$$

Acoustic Wave Speeds

Wave Propagation Direction	Wave Character	Expression for Wave Speed	Wave Speed (in units of 10^5 cm/s)
[001]	V_L (longitudinal)	$(C_{33}/\rho)^{1/2}$	13.1
	V_T (transverse)	$(C_{44}/\rho)^{1/2}$	7.1
[100]	V_L (longitudinal)	$(C_{11}/\rho)^{1/2}$	12.5
	V_T (transverse, polarization along [001])	$(C_{44}/\rho)^{1/2}$	7.1
	V_T (transverse, polarization along [010])	$[(C_{11} - C_{12})/2\rho]^{1/2}$	7.8

For definitions of the crystallographic directions see Appendix 3. For other details see R. Truell et al. (1969).

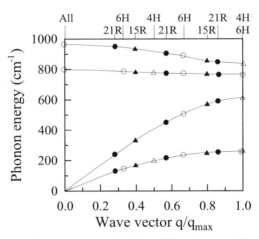

FIG. 5.6.4. Combined dispersion curves for four SiC polytypes: $4H$, $6H$, $15R$, and $21R$ for each polytype, the Raman accessible values of q/q_{max} are marked at the top of the figure [Feldman et al. (1968)].

Main Phonon Energies
(in meV) [Freitas (1995)]

4H-SiC		6H-SiC	
TA_1	46.7	TA_1	46.3
TA_2	51.4	TA_2	53.5
	53.4		
LA	76.9	LA	77.0
	78.8		
TO_1	95.0	TO_1	94.7
		TO_2	95.6
LO	104.0	LO	104.2
	104.3		

Piezoelectric Constants

e_{15}	$0.08 \ C \ m^{-2}$
e_{33}	$0.20 \ C \ m^{-2}$

REFERENCES

Achtziger, N., and W. Witthuhn, *Appl. Phys. Lett.* **71**, 1 (1997), 110–112.

Anikin, M.M., M.E. Levinshtein, I.V. Popov, V.P. Rastegaev, A.M. Strel'chuk, and A.L. Syrkin, *Soviet Phys. Semicond.* **22**, 9 (1988), 995–998.

Anikin, M.M., A.S. Zubrilov, A.A. Lebedev, A.M. Strelchuk, and A.E. Cherenkov, *Soviet Phys. Semicond.* **25**, 3 (1991), 289–293.

Barrett, D.L., and R.B. Campbell, *J. Appl. Phys.* **38**, 1 (1967), 53.

Bellotti, E., H.E. Nilsson, K.F. Brennan, and P.P. Ruden, *J. Appl. Phys.* **85**, 6 (1999), 3211–3217.

Burgemeister, E.A., W. von Muench, and E. Pettenpaul, *J. Appl. Phys.* **50**, 9 (1979), 5790–5794.

Casady, J.B., and R.W. Johnson, *Solid State Electr.* **39**, 10 (1996), 1409–1422.

Chen, G.D., J.Y. Lin, and H.X. Jiang, *Appl. Phys. Lett.* **68**, 10 (1996), 1341–1343.

Chinone, Y., S. Ezaki, F. Fuijita, and R. Matsumoto, *Springer Proc. Phys.* **43** (1989), 198–206.

Choyke, W.J., *Mater. Res. Bull.* **4** (1969), 141.

Choyke, W.J., and G. Pensl, *Mater. Res. Bull* (1997), 25–29.

Dalibor, T., G. Pensl, N. Nordell, A. Schoner, and W.J. Choyke *Abstracts on International Conference on SiC, III–V Nitrides and Related Materials*, Aug. 31–Sept. 5, 1997, Stockholm, Sweden, pp. 55–56.

Dean, P.J., W.J. Choyke, and L. Patric, *J. Lumin.* **10** (1977), 299–314.

Dombrowski, K.F., U. Kaufmann, M. Kunzer, K. Mayer, V. Schneider, V.B. Shields, and M.G. Spencer, *Appl. Phys. Lett.* **65**, 14 (1994), 1811–1813.

Drowart, J., G. De Maria, and M.G. Ingram, *J. Chem. Phys.* **29**, 5 (1958), 1015–1021.

Dubrovskii, G.B., and E.I. Radovanova, *Phys. Lett.* **28A**, 4 (1968), 283–284.

Dubrovskii, G.B., and E.I. Radovanova, *Phys. Stat. Solidi (b)* **48**, 2 (1971), 875–879.

Dubrovskii, G.B., A.A. Lepneva, and E.I. Radovanova, *Phys. Stat. Solidi (b)* **57**, 1 (1973), 423–431.

Evwaraye, A.O., S.R. Smith, and W.C. Mitchel, *J. Appl. Phys.* **75**, 7 (1994), 3472–3476.

Evwaraye, A.O., S.R. Smith, and W.C. Mitchel, *J. Appl. Phys.* **79**, 10 (1996), 7726–7730.

Evwaraye, A.O., S.R. Smith, W.C. Mitchel, and H.McD. Hobgood, *Appl. Phys. Lett.* **71**, 9 (1997), 1186–1188.

Feldman, D.W., J.H. Parker, W.J. Choyke, and L. Patrick, *Phys. Rev.* **173**, 3 (1968), 787–793.

Frantsevich, I.N., F.F. Voronov, and S.A. Bakuta, *Elastic Constants and Elastic Modulus in Metals and Non-metals*, Naukova Dumka, Kiev, (1982), in Russian. (See also *Anisotropy in Single Crystal Refractory Compounds*, Vol. 2, Plenum Press, New York, 1968, p. 493.

Freitas, J.A., Jr. *Photoluminescence Spectra of SiC Polytypes*, in *Properties of Silicon Carbide* (edited by Harris, G.L.), EMIS Datareviews Series, No. 13, 1995, pp. 29–41.

Galeskas, A., J. Linnros, V. Grivickas, U. Lindefelt, and C. Hallin, *Proceedings of the 7th International Conference on SiC, III-Nitrides and Related Materials*, Aug. 31–Sept. 5, 1997, Stockholm, Sweden 1997, pp. 533–536.

Gmelins Handbuch der Anorganischen Chemie, 8th edition, *Silicium*, Part B, Weinheim, Verlag Chemie, GmbH, 1959.

Grivickas, V., J. Linnros, and A. Galeskas, *Proceedings of the 7th International Conference on SiC, III-Nitrides and Related Materials*, Aug. 31–Sept. 5, 1997, Stockholm, Sweden, 1997, pp. 529–532.

Groth, R., and E. Kauer, *Phys. Stat. Solidi* **1**, 5 (1961), 445–450.

Harris, G.L., Thermal conductivity of SiC, in *Properties of Silicon Carbide* (edited by Harris, G.L.), EMIS Datareviews Series, No. 13, 1995a, pp. 5–6.

Harris, G.L., Density of SiC, in *Properties of Silicon Carbide* (edited by Harris, G.L.), EMIS Datareviews Series, No. 13, 1995b, p. 3.

Harris, G.L., Young's modulus of SiC, in *Properties of Silicon Carbide* (edited by Harris, G.L.), EMIS Datareviews Series, No. 13, 1995c, p. 8.

Hobgood, McD.H., R.C., Glass, G. Augustine, R.H. Hopkins, J. Jenny, M. Skowronski, W.C. Mitchel, and M. Roth, *Appl. Phys. Lett.* **66**, 11 (1995), 1364–1366.

Ikeda, M., H. Matsunami, and T. Tanaka, *Phys. Rev. B* **22**, 6 (1980), 2842–2854.

Ivanov, P.A., M.E. Levinshtein, K.G. Irvine, O. Kordina, J.W. Palmour, S.L. Rumyantsev, and R. Singh, *Electron. Lett.* **35**, 11 (1999), 1382–1383.

Kamitani, K., M. Grimsditch, J.C. Nipko, C.-K. Loong, M. Okada, and I. Kimura, *J. Appl. Phys.* **82**, 6 (1997), 3152–3154.

Karch, K., P. Pavone, W. Windl, O. Schutt, and D. Strauch, *Phys. Rev. B* **50**, 23 (1994), 17054–17063.

Kern, E.L., D.W. Hamil, H.W. Deam, and H.D. Sheets, *Mater. Res. Bull., Special Issue* **4** (1969), S25–S32. (*Proceedings of the International Conference on Silicon Carbide*, University Park, Pennsylvania, USA, October 20–23 1968).

Khan, I.A., and J.A. Cooper, *Am. Sci. Forum* **264–268** (1998), 509–512.

Kimoto, T., N. Miyamoto, and H. Matsunami, *IEEE Trans. Electron Dev.* **46**, 3 (1999), 471–477.

Kinoshita, T., M. Schadt, K.M. Itoh, J. Muto, and G. Pensl, *Abstracts of International Conference on Silicon Carbide, III-Nitrides, and Related Materials* Stockholm, Sweden, Aug. 31–Sept. 5, 1997, 631–632.

Konstantinov, A.O., *Soviet Phys. Semicond.* **23**, 1 (1989), 31–35.

Konstantinov, A.O., Q. Wahab, N. Nordell, and U. Lindefelt, *Appl. Phys. Lett.* **71**, 1 (1997), 90–92.

Konstantinov, A.O., N. Nordell, Q. Wanab, and U. Lindefelt, *Appl. Phys. Lett.* **73**, 13 (1998), 1850–1852.

Kordina, O., J.P. Bergman, C. Hallin, and E. Janzen, *Appl. Phys. Lett.* **69**, 5 (1996), 679–681.

Kuwabara, H., and S. Yamada, *Phys. Stat. Solidi* **A30** (1975), 739–746.

Kuwabara, H., K. Yamanaka, and S. Yamada, *Phys. Stat. Solidi* **A37** (1976), K157–K161.

Kuznetsov, N.I., A.P. Dmitriev, and A.S. Furman, *Semiconductors* **28**, 6 (1994), 584–586.

Kuznetsov, N.I., and A.S. Zubrilov, *Mater. Sci. Eng.* **B29** (1995), 181–184.

Kuznetsov, N.I., and J.A. Edmond, *Semiconductors* **31**, 10 (1997), 1049–1052.

Kyuregyan, A.S., and S.N. Yurkov, *Soviet Phys. Semicond.* **23**, 10 (1989), 1126–1132.

Lambrecht, W.R.L., B. Segall, W. Suttrop, M. Yoganathan, R.P. Devaty, W.J. Choyke, J.A. Edmond, J.A. Powell, and M. Alouani, *Appl. Phys. Lett.* **63**, 20 (1993), 2747–2749.

Lambrecht, W.R.L., B. Segall, W. Suttrop, M. Yoganathan, R.P. Devaty, W.J. Choyke, J.A. Edmond, J.A. Powell, and M. Alouani, *Phys. Rev. B* **50**, 15 (1994), 10722–10726.

Lebedev, A.A., and Poletaev, N.K., *Semiconductors* **30** (1996), 238–241.

Lebedev, A.A., and D.V. Davydov, *Semiconductors* **31**, 9 (1997), 935–937.

Lebedev, A.A., *Semiconductors* **33**, 2 (1999), 107–130.

Lindefelt, U., *J. Appl. Phys.* **84**, 5 (1998), 2628–2637.

Logothetidis, S., and J. Petalas, *J. Appl. Phys.* **80**, 3 (1996), 1768–1772.

Marshall, R., *Mater. Res. Bull., Special Issue* **4** (1969), S73–S84. (*Proceedings of the International Conference on Silicon Carbide*, University Park, Pennsylvania, October 20–23, 1968).

Mickevicius, R., and J.H. Zhao, *J. Appl. Phys.* **83**, 6 (1998), 3161–3167.

Mitchel, W.C., R. Perrin, J. Goldstein, M. Roth, S.R. Ahoujja, A.D. Smith, A.O. Evwaraye, J.S. Solomon, G. Laudis, J. Jenny, and H.McD. Hobgood, *Abstracts of*

International Conference on SiC, III-Nitrides and Related Materials, Aug. 31–Sept. 5, 1997, Stockholm, Sweden, 1997, pp. 585–586.

Morelli, D., J. Hermans, C. Beetz, W.S. Woo, G.L. Harris, and C. Taylor, in *Silicon Carbide and Related Materials* (edited by Spencer, M.G., et al.), *Institute of Physics Conference Series* No. 137, pp. 313–316, 1993.

von Muench, W., and E. Pettenpaul, *J. Appl. Phys* **48** (1977), 4823–4825.

Nakashima, S., and K. Tahara, *Phys. Rev. B* **40**, 9 (1989), 6339–6344.

Neudeck, P.G., and C. Fazi, *Proceedings of the 7th Int. Conf. on SiC, III-Nitrides and Relat. Mat.*, Aug. 31–Sept. 5, 1997, Stockholm, Sweden, 1997, pp. 1037–1040.

Nilsson, O., H. Mehling, R. Horn, J. Fricke, R. Hofmann, S.G. Muller, R. Eckstein, and D. Hofmann, *High Temperatures–High Pressures* **29** (1997), 73–79.

Olego, D., and M. Cardona, *Phys. Rev. B* **25**, 2 (1982), 1151–1160.

Park, C.H., B.-H. Cheong, K.-H. Lee, and K.J. Chang, *Phys. Rev. B* **49**, 7 (1994), 4485–4493.

Patrick, L., and W.J. Choyke, *Phys. Rev.* **186**, 3 (1969), 775–777.

Persson, C., and U. Lindefelt, *J. Appl. Phys.* **82**, 11 (1997), 5496–5508.

Philipp, H.R., and E.A. Taft, *Silicon Carbide—a High Temperature Semiconductor* (edited by J.K. O'Connor and J. Smieltens), Pergamon Press, Oxford, 1960, p. 366.

Radovanova, E.I., Ph.D. thesis, The Ioffe Institute of Russian Academy of Science, St. Petersburg, Russia (1973).

Raghunathan, R., and Baliga, B.J., *Appl. Phys. Lett.* **72**, 24 (1998), 3196–3198.

Raghunathan, R., and Baliga, B.J., *Solid State Electronics* **43** (1999), 199–211.

Ruff, M., H. Mitlehner, and R. Helbig, *IEEE Trans. Electron Dev.* **41**, 6 (1994), 1040–1054.

Rutsch, G., R.P. Devaty, W.J. Choyke, D.W. Langer, and L.B. Rowland, *J. Appl. Phys.* **84**, 4 (1998), 2062–2064.

Shaffer, P.T.B., in *Plenum Press Handbook of High Temperature Materials*, No. 1 (1964), Plenum Press., New York, p. 107.

Shaffer, P.T.B., *J. Am. Ceram. Soc.* **48**, 11 (1965), 601.

Shaffer, P.T.B., and R.G. Naum, *J. Opt. Soc. Am.* **59**, 11 (1969), 1498.

Shaffer, P.T.B., *Appl. Opt.* **10** (1971), 1034–1036.

Shaffer, W.J., H.S. Kong, G.H. Negley, and J.W. Palmour, *Inst. Phys. Conf. Ser.* **137** (1994), 155.

Shinohara, M., M. Yamanaka, H. Daimon, E. Sakuma, H. Okumura, S. Misawa, K. Endo, and S. Yoshida, *Jpn. J. Appl. Phys.* **27**, 3 (1988), L433–L436.

Solangi, A., and M.I. Chaudry, *J. Mater. Res.* **7** (1992), 539–541.

Son, N.T., O. Kordina, A.O. Konstantinov, W.M. Chen, E. Sorman, B. Monemar, and E. Janzen, *Appl. Phys. Lett.* **65**, 25 (1994), 3209–3211.

Son, N.T., W.M. Chen, O. Kordina, A.O. Konstantinov, B. Monemar, E. Janzen, D.M. Hofman, D. Volm, M. Drechsler, and B.K. Meyer, *Appl. Phys. Lett.* **66**, 9 (1995), 1074–1076.

Spitzer, W.J., D.A. Kleinman, and D.J. Walsh, *Phys. Rev.* **113** (1959), 127.

Sridhara, S.G., R.P. Devaty, and W.J. Choyke, *J. Appl. Phys.* **84**, 5 (1998), 2963–2964.

Tairov, Yu.M., and V.F. Tsvetkov, in *Handbook on Electrotechnical Materials* (edited by Koritskii, Yu.V., V.V. Pasynkov, and B.M. Tareev), in 3 volumes, Section 19, pp. 446–471, "*Semiconductor Compounds $A^{IV}\,B^{IV}$,*" Energomashizdat, Leningrad, 1988.

Taylor, A., and R.M. Jones, in *Silicon Carbide—A High Temperature Semiconductor*, Pergamon Press, New York (1960).

Troffer, T., G. Pensl, A. Schoner, A. Henry, C. Hallin, O. Kordina, and E. Jansen, *Abstracts of International Conference on SiC, III-Nitrides and Related Materials*, Aug. 31–Sept. 5, 1997, Stockholm, Sweden, 1997, pp. 601–602.

Truell, R., C. Elbaum, and B.B. Chick, *Ultrasonic Methods in Solid State Physics*, Academic Press, London, 1969.

Kimoto, T., N. Miyamoto, and H. Matsunami, *IEEE Trans. Electron Dev.* **46**, 3 (1999), 471–477.

Van Daal, H.J., W.F. Knippenberg, and J.D. Wasscher, *J. Phys. Chem. Solids* **24** (1963), 109–127.

Yamanaka, M., H. Daimon, E. Sakuma, S. Misawa, and S. Yoshida, *J. Appl. Phys.* **61**, 2 (1987a), 599–603.

Yamanaka, M., H.H. Daimon, E. Sakuma, S. Misawa, K. Endo, and S. Yoshida, *Jpn. J. Appl. Phys.* **26**, 5 (1987b), L533–L535.

Chapter 6

Silicon-Germanium (Si$_{1-x}$Ge$_x$)

F. Schäffler

Johannes Kepler University, Linz, Austria

The group IV silicon-germanium random alloys differ in several respects from other material combinations treated in this volume. One of the most characteristic features of this material combination concerns bulk Si$_{1-x}$Ge$_x$: Si and Ge are miscible over the complete range of compositions. However, the large splitting of the solidus/liquidus phase boundary makes it almost impossible to pull bulk crystals of acceptable radial and axial homogeneity in a composition range that differs from the pure materials by more than a few atomic percent. Hence, interest in bulk alloys, which undoubtedly existed in the 1960s and early 1970s for a variety of reasons, soon waned both because of the rapid switching from Ge to Si as the dominant device material and because of the fundamental difficulties of providing high-quality Si$_{1-x}$Ge$_x$ substrates. For these reasons, many of the available data concerning the physical properties of bulk Si$_{1-x}$Ge$_x$ were recorded more than 30 years ago, some of them on material of doubtful crystal quality, especially in the composition range around $x = 50\%$. Material quality and composition-dependent dopant segregation are also the main reason for the almost complete lack of data concerning the dopant-dependence of basic material

Properties of Advanced Semiconductor Materials, Edited by Levinshtein, Rumyantsev, and Shur.
ISBN 0-471-35827-4 © 2001 John Wiley & Sons, Inc.

parameters, such as carrier mobility or fundamental energy gap. Except for a few new attempts toward employing Si$_{1-x}$Ge$_x$ bulk alloys with moderate compositions for thermoelectric and optoelectronic devices, only minor activities became known in that direction in the last 20 years or so.

On the other hand, the development of low-temperature growth techniques, such as molecular beam epitaxy or chemical vapor deposition, and the new concepts of energy band engineering (first emerged for III–V materials in the seventies and eighties) led to a fast increase of the number of groups dealing with thin Si$_{1-x}$Ge$_x$ films. These films are usually deposited on an Si substrate. Because of the inherent lattice mismatch of around 4% between pure Si and pure Ge, such films are tetragonally distorted, when grown to a thickness below the critical value for the onset of misfit dislocations. These films begin to relax to their intrinsic cubic lattice constant, once the critical thickness is exceeded. Hence depending on the thickness of a Si$_{1-x}$Ge$_x$ film at a given composition (and other growth parameters that rule kinetic limitations), such films can be either biaxially strained or strain-relaxed. Strain-relaxed films can be considered as sort of a virtual "bulk" substrate. With the art of epitaxial growth rapidly advancing, it was soon recognized that strain is an as important material parameter as composition in the Si/Si$_{1-x}$Ge$_x$ heterostructure system. Many parameters, such as band gaps, band offsets, effective masses, and so on, are strongly strain-dependent, making strain control a vital necessity for any kind of energy band engineering conceivable in these materials. The advantages gained by the introduction of thin Si$_{1-x}$Ge$_x$ films in their basic compatibility with standard silicon technologies have made this heterostructure system an extremely interesting candidate for production devices. The first commercial products in the high-frequency analog market segment were introduced in spring 1998.

This chapter reflects the contrast between the technical relevance of strained Si$_{1-x}$Ge$_x$ thin films and the quite limited interest in bulk alloys. Where strain-dependent data are given, they are restricted to biaxial strain in the (001) plane, which corresponds to pseudomorphic growth on a (001)-oriented substrate. This is presently the only orientation of technical relevance, but references to other surface orientations are given, when available. Also, data that are important for the relaxation of Si$_{1-x}$Ge$_x$ on Si, such as critical thickness or misfit dislocation glide velocities, are incorporated. The more detailed part of this chapter is preceded by a table of 300 K bulk data, which also repeats the most basic properties of elemental Si and Ge for comparison. Several parameters, such as the lattice constant, vary almost linearly between the constituents, expressing their close chemical similarity. Other parameters, such as the band gap or the effective electron masses, do not, because the general conduction band structure changes from Si-like to Ge-like at $x = 85\%$. The variation of the room temperature bulk parameters with composition, as listed in the table, gives a quick guide to where linear variation can be expected and where not. In any case, the more detailed plots in the second part should be consulted, especially when dealing with strained films of $Si_{1-x}Ge_x$ alloys. Additional data and physical background can be found in the well-known Landoldt-Börnstein Series [Landoldt-Börnstein (1982, 1987) and in Vol. 12 of the EMIS Datareview Series [Kasper (1995)].

6.1. BASIC PARAMETERS IN UNSTRAINED BULK MATERIAL AT 300 K

	Si	Ge	$Si_{1-x}Ge_x$
Crystal structure	Diamond	Diamond	Diamond (random alloy)
Group of symmetry	$O_h^7 - Fd3m$	$O_h^7 - Fd3m$	$O_h^7 - Fd3m$
Number of atoms (cm^{-3})	5.00×10^{22}	4.42×10^{22}	$(5.00 - 0.58x)$ $\times 10^{22}$
Debye temperature (K)	640	374	$640 - 266x$
Density (g/cm^3)	2.329	5.323	$2.329 + 3.493x$ $- 0.499x^2$
Dielectric constant	11.7	16.2	$11.7 + 4.5x$
Effective electron mass (in units of m_0)			
longitudinal m_l	0.92	1.59	~ 0.92 for $x < 0.85$ ~ 1.59 for $x > 0.85$

(Continued)

	Si	Ge	Si$_{1-x}$Ge$_x$
transversal m_t	0.19	0.08	~0.19 for $x < 0.85$
			~0.08 for $x > 0.85$
Effective hole mass (in units of m_0)			
heavy m_{hh}	0.54	0.33	
light m_{lh}	0.15	0.043	
spin–orbit–split m_{so}	0.23	0.095	$0.23 - 0.135x$
Electron affinity (eV)	4.05	4.0	$4.05 - 0.05x$
Lattice constant (Å)	5.431	5.658	$5.431 + 0.2x + 0.027x^2$
Optical phonon energy (meV)			
$\hbar\omega$ Si–Si	63		$63 - 8.7x$
$\hbar\omega$ Ge–Ge		37	$35 + 2.0x$
$\hbar\omega$ Si–Ge			≈ 50 (see Fig. 6.6.9)

Band structure and carrier concentration

	Si	Ge	Si$_{1-x}$Ge$_x$
Energy gap (eV) (indirect)			
Δ conduction band min.	1.12		$1.12 - 0.41x + 0.008x^2$ $(x < 0.85)$
L conduction band min.		0.66	$1.86 - 1.2x$ $(x > 0.85)$
Direct energy gaps (eV):			
Energy separations $E_{\Gamma 1}$	3.4	0.8	see Figs. 6.2.1 and 6.2.2
Energy separations $E_{\Gamma 2}$	4.2	3.2	see Figs. 6.2.1 and 6.2.2
Spin–orbit splitting (eV)	0.044	0.29	$0.044 + 0.246x$
Intrinsic carrier concentration (cm^{-3})	1×10^{10}	2×10^{13}	see Fig. 6.2.7

(Continued)

	Si	Ge	$Si_{1-x}Ge_x$
Effective conduction band density of states (cm^{-3})	2.8×10^{19}	1.0×10^{19}	$\sim 2.8 \times 10^{19}$ $(x < 0.85)$ $\sim 1.0 \times 10^{19}$ $(x > 0.85)$
Effective valence band density of states (cm^{-3})	1.8×10^{19}	5.0×10^{18}	
Electrical properties			
Breakdown field (V/cm)	3×10^5	10^5	$<3 \times 10^5$
Mobility $(cm^2/V \cdot s)$			
electrons	1450	3900	$1450{-}4325x$ $(0 \le x < 0.3)$
holes	450	1900	$450{-}865x$ $(0 \le x < 0.3)$
Diffusion coefficient (cm^2/s)			
electrons	36	100	$36{-}112x$ $(0 \le x < 0.3)$
holes	12	50	$12{-}22x$ $(0 \le x < 0.3)$
Thermal velocity (cm/s)			
electrons	2.4×10^7	3.1×10^7	$\approx 2.4 \times 10^7$ $(x < 0.85)$ $\approx 3.1 \times 10^7$ $(x > 0.85)$
holes	1.65×10^7	1.9×10^7	$1.65 + 0.25x$
Optical properties			
Index of refraction	3.42	4.0	$3.42 + 0.37x$ $+ 0.22x^2$
Radiative recombination coefficient (cm^3/s)	1.1×10^{-14}	6.4×10^{-14}	

(Continued)

	Si	Ge	$Si_{1-x}Ge_x$
Thermal and mechanical properties			
Elastic moduli			
c_{11} (GPa)	165.8	128.5	$165.8 - 37.3x$
c_{12} (GPa)	63.9	48.3	$63.9 - 15.6x$
c_{44} (GPa)	79.6	66.8	$79.6 - 12.8x$
Bulk modulus (GPa)	98	75	$98 - 23x$
Poisson ratio σ_{100}	0.28	0.26	$0.28 - 0.02x$
Melting point (°C)	1412	937	$1412 - 738x + 263x^2$ (solidus)
			$1412 - 80x - 395x^2$ (liquidus)
Specific heat			
J/mol K	19.6	22.5	$19.6 + 2.9x$
J/g K	0.7	0.31	
Thermal conductivity (W/cm K)	1.3	0.58	$\sim 0.046 + 0.084x$ $(0.2 < x < 0.85)$
Thermal diffusivity (cm^2/s)	0.8	0.36	—
Linear thermal expansion (1/K)	2.6×10^{-6}	5.9×10^{-6}	$(2.6 + 2.55x) \times 10^{-6}$ $(x < 0.85)$
			$(7.53x - 0.89) \times 10^{-6}$ $(x > 0.85)$

6.2. BAND STRUCTURE AND CARRIER CONCENTRATION

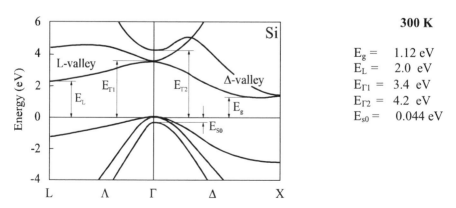

FIG. 6.2.1. Band structure of Si. Important minima of the conduction band and maxima of the valence band.

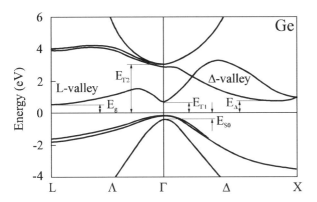

FIG. 6.2.2. Band structure of Ge. Important minima of the conduction band and maxima of the valence band.

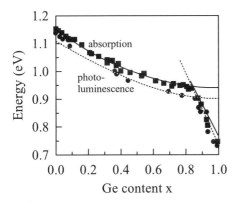

FIG. 6.2.3 Fundamental (indirect) band gap of $Si_{1-x}Ge_x$ (absorption measurements: squares) and excitonic band gap (photoluminescence measurements: dots) at 4.2 K [Braunstein et al. (1958) and Weber and Alonso (1989)].

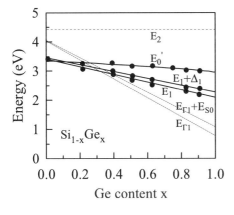

FIG. 6.2.4. Composition dependences of several important direct transitions observed in $Si_{1-x}Ge_x$. For details see Kline et al. (1968) and Pickering et al. (1993).

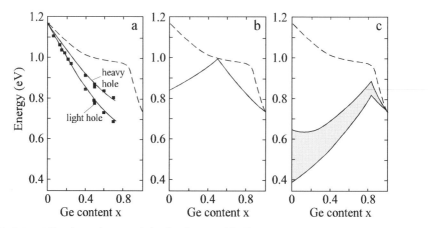

FIG. 6.2.5. The dependences of the fundamental indirect band gap versus x of pseudo-morphic Si$_{1-x}$Ge$_x$ (001) alloys: (a) On Si substrate; (b) on Si$_{0.5}$Ge$_{0.5}$ substrate; and (c) on Ge substrate. Dashed lines represent the dependences for unstrained bulk band gap. Experimental points are taken from Lang et al. (1985) and Dutartre et al. (1991). Solid lines are the calculated curves from People (1985, 1986) and Van de Walle and Martin (1986).

Linear Fitting of Direct Energy Gaps Versus Composition

Electroreflectance experiments on *strained* **Si$_{1-x}$Ge$_x$ films** on Si substrates at 70 K [Ebner et al. (1998)]:

$$0 < x < 0.3$$

$$E_{\Gamma 1}\,(\text{meV}) = 4175 - (2814 \pm 55)x$$

$$E_{\Gamma 2}\,(\text{meV}) = 3400 \pm 7 - (300 \pm 40)x$$

$$E_1\,(\text{meV}) = 3452 - (1345 \pm 25)x$$

$$E_1'\,(\text{meV}) = 5402 \pm 25 + (280 \pm 120)x$$

$$E_2(X)\,(\text{meV}) = 4351 \pm 38 + (210 \pm 180)x$$

$$E_2(\Sigma)\,(\text{meV}) = 4518 \pm 129 + (880 \pm 600)x$$

6.2.1. Temperature Dependences

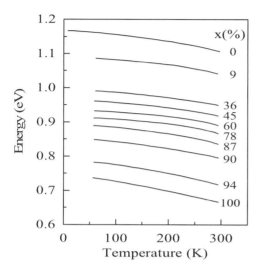

FIG. 6.2.6. Temperature dependence of the fundamental indirect band gaps of $Si_{1-x}Ge_x$ alloys at different x [Braunstein et al. (1958)].

Temperature Dependence of the Energy Gap

$x = 0$ (Si):

$$E_g = 1.17 - 4.73 \times 10^{-4} \times \frac{T^2}{T + 636} \text{ (eV)} \tag{6.2.1}$$

$x = 1$ (Ge):

$$E_g = 0.742 - 4.8 \times 10^{-4} \times \frac{T^2}{T + 235} \text{ (eV)} \tag{6.2.2}$$

Temperature Dependence of the Direct Band Gap

$x = 0$ (Si):

$$E_{\Gamma 2} = 4.34 - 3.91 \times 10^{-4} \times \frac{T^2}{T + 125} \text{ (eV)} \tag{6.2.3}$$

$x = 0$ (Si):

$$E_{\Gamma 1} = 0.89 - 5.82 \times 10^{-4} \times \frac{T^2}{T + 296} \text{ (eV)} \tag{6.2.4}$$

where T is temperature in degrees K.

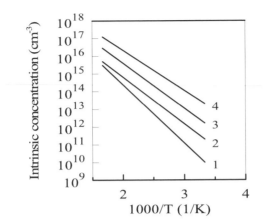

FIG. 6.2.7. The temperature dependences of the intrinsic carrier concentration of Si$_{1-x}$Ge$_x$ alloys at different x. Curve 1, $x = 0$ (Si); curve 2, $x = 0.4$; curve 3, $x = 0.8$; curve 4, $x = 1.0$ (Ge).

Intrinsic carrier concentration:

$$n_i = (N_c \cdot N_v)^{1/2} \exp\left(-\frac{E_g}{2k_B T}\right) \qquad (6.2.5)$$

Effective Density of States in the Conduction Band N_c

At $x < 0.85$, Si$_{1-x}$Ge$_x$ alloys are considered as "Si-like" material:

$$N_c \cong 4.82 \times 10^{15} \cdot M\left(\frac{m_c}{m_0}\right)^{3/2} \cdot T^{3/2} \ (\text{cm}^{-3}) \cong$$

$$N_c \cong 4.82 \times 10^{15} \cdot \left(\frac{m_{cd}}{m_0}\right)^{3/2} \cdot T^{3/2} \ (\text{cm}^{-3})$$

$$\cong 5.3 \times 10^{15} \times T^{3/2} \ (\text{cm}^{-3}) \qquad (6.2.6)$$

$M = 6$ is the number of equivalent valleys in the conduction band.
$m_c = 0.32m_0$ is the effective mass of density of states in one valley of conduction band.
$m_{cd} = 1.06m_0$ is the effective mass of density of states.

At $0.85 < x < 1$, $Si_{1-x}Ge_x$ alloys are considered as "Ge-like" material:

$$N_c \cong 4.82 \times 10^{15} \cdot M \left(\frac{m_c}{m_0}\right)^{3/2} \cdot T^{3/2} \ (cm^{-3}) \cong$$

$$N_c \cong 4.82 \times 10^{15} \cdot \left(\frac{m_{cd}}{m_0}\right)^{3/2} \cdot T^{3/2} \ (cm^{-3})$$

$$\cong 2 \times 10^{15} \times T^{3/2} \ (cm^{-3}) \tag{6.2.7}$$

$M = 4$ is the number of equivalent valleys in the conduction band. $m_c = 0.22m_0$ is the effective mass of density of states in one valley of conduction band.

$m_{cd} = 0.55m_0$ is the effective mass of density of states.

Effective Density of States in the Valence Band N_V

$x = 0$ (Si):

$$N_v \cong 4.82 \times 10^{15} \cdot \left(\frac{m_v}{m_0}\right)^{3/2} \cdot T^{3/2} \ (cm^{-3})$$

$$\cong 3.5 \times 10^{15} \times T^{3/2} \ (cm^{-3}) \tag{6.2.8}$$

where $m_v = 0.81m_0$ is the hole effective mass of density of states.

$x = 1$ (Ge):

$$N_v \cong 4.82 \times 10^{15} \cdot \left(\frac{m_v}{m_0}\right)^{3/2} \cdot T^{3/2} \ (cm^{-3})$$

$$\cong 9.6 \times 10^{14} \times T^{3/2} \ (cm^{-3}) \tag{6.2.9}$$

where $m_v = 0.34m_0$ is the hole effective mass of density of states.

There is a large uncertainty in the available data on the composition dependence of hole effective mass of density of states m_v. For crude estimations one can use the simplest linear approximation:

$$m_v(x) \cong (0.81 - 0.47x)m_0 \tag{6.2.10}$$

(see also Section 6.2.4).

6.2.2. Dependence of Energy Gap on Hydrostatic Pressure

$x = 0$ (Si):

$$E_g = E_g(0) - 1.4 \times 10^{-3}P \text{ (eV)} \qquad (6.2.11)$$

$x = 1$ (Ge):

$$E_g = E_g(0) + 5.1 \times 10^{-3}P \text{ (eV)} \qquad (6.2.12)$$

where P is a pressure in kbar.

6.2.3. Strain-Dependent Band Discontinuity

Band discontinuities at an Si$_{1-x}$Ge$_x$/Si$_{1-y}$Ge$_y$ are only defined, if the interface is coherent—that is, if the in-plane lattice constant is preserved across the interface. Generally, both layers are biaxially strained in the interface plane. The biaxial strain status can be converted into a hydrostatic and a uniaxial component, whose sign is opposite to that of the in-plane strain and directed perpendicular to the interface. Both the interface chemistry and the biaxial strain status are relevant for the determination of the band discontinuities at the interface, which will be given in the following for growth on a (001) surface.

The chemical contribution gives a linear variation of the weighted average valence band discontinuity ΔE_v between a strained Si$_{1-x}$Ge$_x$ film and an unstrained cubic Si$_{1-xs}$Ge$_{xs}$ substrate [Rieger and Vogl (1993)]:

$$\Delta E_v(x, xs) = (0.47 - 0.06x)(x - xs) \text{ (eV)} \qquad (6.2.13)$$

The hydrostatic strain components leads to a change in the band gap of the strained Si$_{1-x}$Ge$_x$ film according to

$$\Delta E_g = (\Xi_d + 1/3\Xi_u - a)(\varepsilon_\perp + 2\varepsilon_\parallel) \qquad (6.2.14)$$

where
$(\Xi_d + 1/3\Xi_u - a)$ is the deformation potential difference relevant for the fundamental band gap (see Table I).

$\varepsilon_\perp = (a_\perp - a_0)/a_0$, $\varepsilon_\parallel = (a_\parallel - a_0)/a_0$ are the perpendicular and in-plane strain components of the strained $Si_{1-x}Ge_x$ film.

a_0 is the undistorted lattice constant of the film.

$a_\perp$ and $a_\parallel$ are the respective components in the strained film.

Both the valence and conduction band degeneracy are lifted by the uniaxial [001] strain component, which leads to the following splittings (Van de Walle and Martin, (1986)):

Valence Band

Heavy hole:

$$E_{v2} = 1/3\Delta_0 - 1/2\delta E_{001} \tag{6.2.15}$$

Light hole:

$$E_{v1} = -1/6\Delta_0 + 1/4\delta E_{001} + 1/2(\Delta_0{}^2 + \Delta_0\delta E_{001} + 9/4\delta E_{001}^2)^{1/2} \tag{6.2.16}$$

Spin–split hole:

$$E_{v3} = -1/6\Delta_0 + 1/4\delta E_{001} - 1/2(\Delta_0{}^2 + \Delta_0\delta E_{001} + 9/4\delta E_{001}^2)^{1/2} \tag{6.2.17}$$

with $\delta E_{001} = 2b(\varepsilon_\perp - \varepsilon_\parallel)$, b being a valence band deformation potential (Table I)

Conduction Band

$$\Delta E_c(\Delta_2) = +2/3\Xi_u^\Delta(\varepsilon_\perp - \varepsilon_\parallel) \tag{6.2.18}$$

$$\Delta E_c(\Delta_4) = -1/3\Xi_u^\Delta(\varepsilon_\perp - \varepsilon_\parallel) \tag{6.2.19}$$

where

Δ_2 are the two electron valleys along the [001] growth direction.

Δ_4 are the four in-plane electron valleys.

Ξ_u^Δ is the relevant conduction band deformation potential for the Δ electrons that define the conduction band minimum in Si-like $Si_{1-x}Ge_x$, $(x \le 0.85)$.

For higher Ge contents, the conduction band becomes Ge-like with electrons being located at the L minimum. With the uniaxial strain component being directed along [001], no splitting of the L minimum occurs for reasons of symmetry.

Table I. Deformation Potentials for the Calculation of the Band Discontinuities[a]

	Si		Ge	
	Theory	Experiment	Theory	Experiment
$(\Xi_d + 1/3\Xi_u - a)$ (eV) (for Δ valley)	1.72	1.5 ± 0.3	1.31	
$(\Xi_d + 1/3\Xi_u - a)$ (eV) (for L valley)	-3.12		-2.78	-2.0 ± 0.5
b (eV)	-2.35	-2.10 ± 0.1	-2.55	-2.86 ± 0.15
Ξ_u (eV)	9.16	-4.85 ± 0.15	9.42	
Ξ_u (eV)	16.14		15.13	16.2 ± 0.4

[a]Theoretical values are taken from Van de Walle and Martin (1986). Experimental data are taken from Laude et al. (1971), Chandrasekar and Pollak (1977), and Balslev (1966). Lacking predictions or measurements for Si$_{1-x}$Ge$_x$ alloys, a linear interpolation is suggested, which is certainly a compromise, whenever the conduction band changes from Si-like to Ge-like.

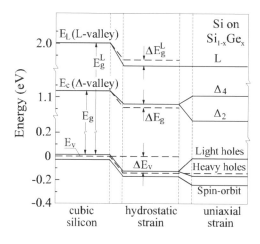

FIG. 6.2.8. Schematic diagram of the relevant band edges of Si subjected to hydrostatic and uniaxial strain as described in equations (6.2.13)–(6.2.19). Energy values apply to a tensely strained Si quantum well on an Si$_{1-x}$Ge$_x$ substrate with $x = 30\%$ [Schaffler (1997)].

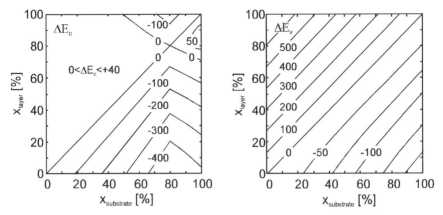

FIG. 6.2.9. Contour plots of the conduction ΔE_c and valence ΔE_v band offsets of pseudomorphic $Si_{1-x}Ge_x$ layers on cubic $Si_{1-xs}Ge_{xs}$ substrates over the complete range of x and xs. The signs correspond to an electronic energy scale, where the active layer (x) is referred to the cubic substrate of composition xs. Exciton-corrected experimental results indicate that for $x > xs$ and $x < 0.8$, the conduction band offset is $0 < \Delta E_c < +40$ meV [Penn et al. (1999)]; that is, for most of the (x, xs) combinations the band alignment is staggered (Type II) with the valence band offset being always in favor of the material with the higher Ge content. The theoretically predicted Type I region for x and xs being larger than about 80% has not been confirmed experimentally as yet [Schäffler (1997)].

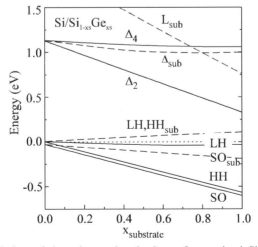

FIG. 6.2.10. Variation of the relevant band edges of a strained Si layer on a cubic $Si_{1-xs}Ge_{xs}$ substrate (solid lines). The dashed lines correspond to the substrate bands. LH, light holes; HH, heavy holes; SO, spin–orbit split holes [Schäffler (1997)].

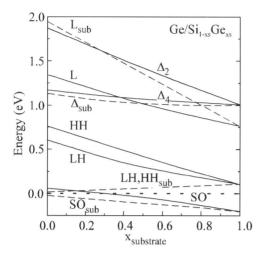

FIG. 6.2.11. Variation of the relevant band edges of a strained Ge layer on a cubic Si$_{1-xs}$Ge$_{xs}$ substrate (solid lines). The dashed lines correspond to the substrate bands [Schaffler (1997)].

6.2.4. Effective Masses

Electrons

Both the Δ and L electron masses are almost unaffected by either composition or biaxial (001) strain [Rieger and Vogl (1993)]. Hence, the electron effective masses are close to the Si bulk value for $x \leq 0.85$ and are close to the Ge bulk values for $x > 0.85$:

At $x < 0.85$, Si$_{1-x}$Ge$_x$ alloys are considered as "Si-like" material:

The surfaces of equal energy are ellipsoids:
 Longitudinal effective mass $m_l = 0.92m_0$
 Transversal effective mass $m_t = 0.19m_0$
Effective mass of density of states in one valley of conduction band:

$$m_c = (m_l \times m_t^2)^{1/3} \approx 0.32m_0$$

Effective mass of conductivity m_{cc}:

$$m_{cc} = 3 \times \left(\frac{1}{m_l} + \frac{2}{m_t} \right)^{-1} \approx 0.26m_0$$

There are $M = 6$ equivalent valleys in the conduction band.
 Effective mass of density of states for all valleys of conduction band:

$$m_{cd} = (M)^{2/3} \times m_c = 1.06m_0$$

At $0.85 < x < 1$, $Si_{1-x}Ge_x$ alloys are considered as "Ge-like" material:

The surfaces of equal energy are ellipsoids:
  Longitudinal effective mass $m_l = 1.59m_0$
  Transversal effective mass $m_t = 0.08m_0$
Effective mass of density of states in one valley of conduction band:

$$m_c = 0.22m_0$$

Effective mass of conductivity m_{cc}:

$$m_{cc} = 3 \times \left(\frac{1}{m_l} + \frac{2}{m_t} \right)^{-1} \approx 0.12m_0$$

There are $M = 4$ equivalent valleys in the conduction band.
  Effective mass of density of states for all valleys of conduction band:

$$m_{cd} = (M)^{2/3} \times m_c = 0.55m_0$$

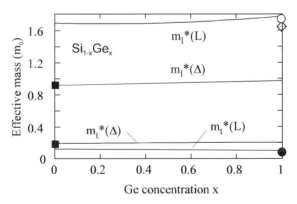

FIG. 6.2.12. Variation of the conduction band effective masses versus composition [Rieger and Vogl (1993)].

Holes

Due to the spin–orbit interaction, hole effective masses depend strongly on composition and crystal direction (warping), and also on strain.

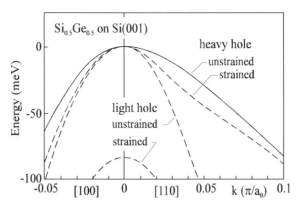

FIG. 6.2.13. A schematic view of the valence band dispersion along [100] and [110] for Si$_{0.5}$Ge$_{0.5}$ on Si(001).

For unstrained bulk, the dispersion can be expressed by three valence band parameters A, B, C [or equivalent representations, Landoldt-Börnstein (1982)] according to

$$E_{hh,lh} = E_v - \frac{\hbar^2 k^2}{2m_0}\left(A \pm \sqrt{B^2 + \frac{C^2}{k^4}(k_x^2 k_y^2 + k_x^2 k_z^2 + k_y^2 k_z^2)}\right) \quad (6.2.20)$$

$$E_{so} = E_v - \Delta - \frac{\hbar^2 k^2}{2m_0}A \quad (6.2.21)$$

where Δ is the spin–orbit splitting (44 meV in Si, 290 meV in Ge), and $k = (k_x, k_y, k_z)$ is the direction in reciprocal space.

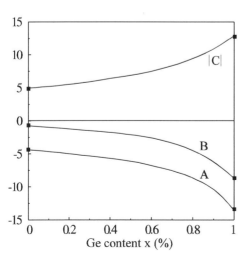

FIG. 6.2.14. Variation of the valence band parameters A, B, and $|C|$ with composition [Schaffler (1997)].

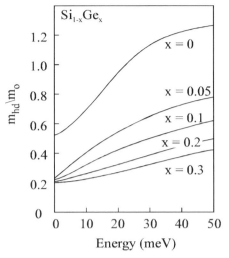

FIG. 6.2.15. The energy dependences of heavy hole effective mass density of states m_{hd} at different x [Manku and Nathan (1991)].

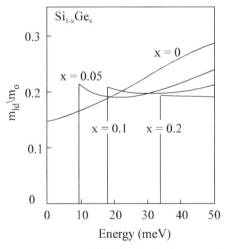

FIG. 6.2.16. The energy dependences of light hole effective mass density of states m_{ld} at different x [Manku and Nathan (1991)].

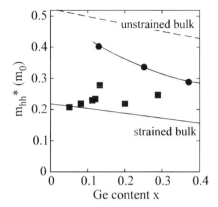

FIG. 6.2.17. Experimental heavy hole cyclotron masses in strained Si$_{1-x}$Ge$_x$ quantum wells. Dots are data collected by Cheng et al. (1994), squares are data from Wong et al. (1995). The dashed line corresponds to unstrained bulk, the bottom solid line is a prediction for strained Si$_{1-x}$Ge$_x$.

6.3. ELECTRICAL PROPERTIES

6.3.1. Mobility and Hall Effect

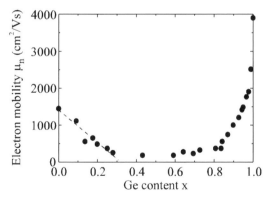

FIG. 6.3.1. Electron Hall mobility versus composition at 300 K [Landolt-Börnstein (1982, 1987), Glicksman (1956)].

At $0 < x < 0.3$

$$\mu_n \approx 1396 - 4315x \ (\text{cm}^2/\text{Vs})$$

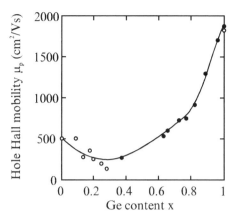

FIG. 6.3.2. Hole Hall mobility versus composition at 300 K [Busch and Vogt (1960), open symbols; Braunstein (1963), full symbols]. Polycrystalline samples were employed in the range $0.3 < x < 0.8$.

For data on Si and Ge, see Levinshtein et al. (1996).

6.3.2. Two-Dimensional Electron Gas

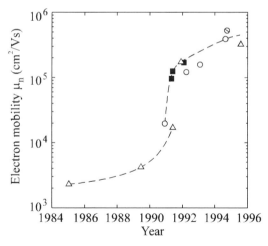

FIG. 6.3.3. Evolution of the published low-temperature electron Hall mobilities. Two-dimensional electron gas in modulation-doped strained Si quantum wells on relaxed $Si_{1-xs}Ge_{xs}$ barriers ($xs \approx 30\%$) [Schaffler (1997)].

The realization of a two-dimensional electron gas in an $Si/Si_{1-x}Ge_x$ heterostructure requires a tensilely inplane strained quantum well for an efficient confinement of the electrons. To suppress alloy scattering, a strained Si quantum well on a cubic Si$_{1-x}$Ge$_x$ buffer layer with x around 30% is usually employed.

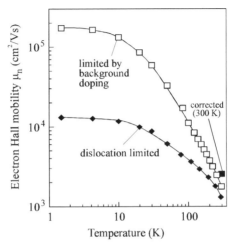

FIG. 6.3.4. Electron Hall mobility versus temperature of strained Si quantum wells on relaxed Si$_{0.7}$Ge$_{0.3}$. For lower curve the mobility is limited by threading dislocations originating from the Si$_{1-x}$Ge$_x$ buffer layer. Background doping limits the mobility of the upper curve. Sheet carrier densities for both curves are 7×10^{11} cm^{-2}. Corrected room temperature mobility of the two-dimensional carriers is 2500 cm^2/V $\cdot$ s [Schäffler (1997)].

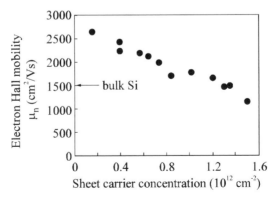

FIG. 6.3.5. Experimental room-temperature Hall mobilities versus measured carrier density. For comparison, the 300 K mobility of undoped bulk Si is marked [Nelson et al. (1993)].

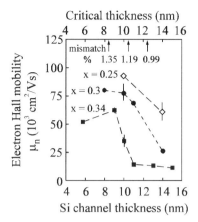

FIG. 6.3.6. Electron Hall mobility at 20 K versus Si channel thickness for modulation-doped Si/Si$_{0.7}$Ge$_{0.3}$ heterostructures. The mobility drop beyond a channel of 100 Å is caused by misfit dislocation formation in the channel [Ismail et al. (1994)].

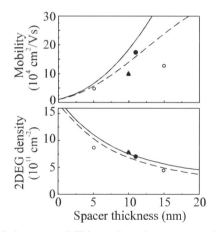

FIG. 6.3.7. Calculated electron mobilities and carrier concentrations versus thickness of the undoped spacer layer between the doping layer and the quantum well. Curves for acceptor background doping levels of 10^{14} cm^{-3} (dashed lines) and 10^{15} cm^{-3} are shown. $T = 1.5$ K. $\Delta E_c = 180$ meV. $N_d = 2 \times 10^{18}$ cm^{-3} [Stern and Laux (1992)].

6.3.3. Two-Dimensional Hole Gas

Two-dimensional hole gases can be realized in pseudomorphic Si$_{1-x}$Ge$_x$ layers on a Si substrate. The experimental mobilities, however, remain far behind theoretical predictions.

An alternative route is the realization of pure Ge channels on a relaxed Si$_{1-x}$Ge$_x$ buffer with $0.6 < x < 0.8$, or of Ge-rich Si$_{1-x}$Ge$_x$ channels ($x < 0.7$) on relaxed Si$_{1-x}$Ge$_x$ buffers with $0.3 < x < 0.5$. Room temperature mobilities of around 1000 cm^2/Vs where reported by Arafa et al. (1996) for such a configuration with a Si$_{0.3}$Ge$_{0.7}$ channel.

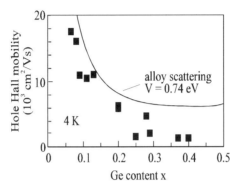

FIG. 6.3.8. Low-temperature hole Hall mobilities versus Ge content in pseudomorphic Si$_{1-x}$Ge$_x$ quantum wells on Si substrates. Data points are taken from Whall (1995). The solid line is a calculated curve for alloy scattering only, using a scattering potential of 0.74 eV [Schaffler (1997)].

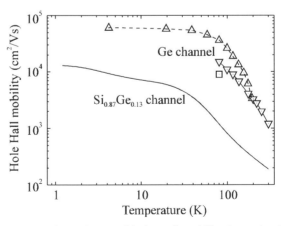

FIG. 6.3.9. Temperature dependences of hole Hall mobility in strained Ge channel on cubic Si$_{0.3}$Ge$_{0.7}$ buffer and in pseudomorphic Si$_{0.87}$Ge$_{0.13}$ quantum well [Schaffler (1997)].

6.4. OPTICAL PROPERTIES

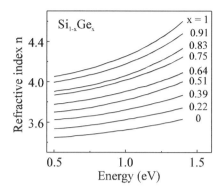

FIG. 6.4.1. Refractive index n in the energy range between 0.5 and 1.4 eV for bulk $Si_{1-x}Ge_x$ [Humlicek (1995)].

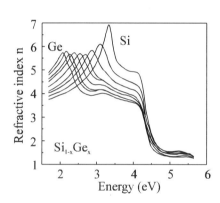

FIG. 6.4.2. Refractive index n in the energy range between 1.5 and 5.5 eV for bulk $Si_{1-x}Ge_x$ alloys. Lines correspond (left to right) to composition values of $x = 1$ (Ge), 0.915, 0.831, 0.75, 0.635, 0.513, 0.389, 0.218, 0 (Si) [Humlicek et al. (1989)].

Infrared refractive index n_∞ at $T = 300$ K:

$$n_\infty \approx 3.42 + 0.37x + 0.22x^2$$

For Si ($x = 0$):

$$\text{At } 77 < T < 400 \text{ K}, \qquad n_\infty = 3.38 \times (1 + 3.9 \times 10^{-5} \times T)$$

$$\text{At } T = 300 \text{ K}, \qquad n_\infty = 3.42$$

For Ge ($x = 1$):

$$\text{At } T = 300 \text{ K}, \qquad n_\infty = 4.00$$

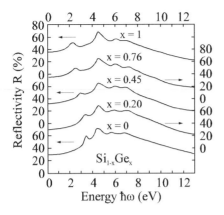

FIG. 6.4.3. Reflectivity versus photon energy in the energy range 0–13 eV [Schmidt (1968)].

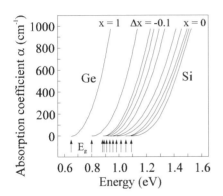

FIG. 6.4.4. Absorption coefficient α in the energy range between 0.5 and 1.4 eV for bulk Si$_{1-x}$Ge$_x$ alloys with x varying by increments of 0.1 between Ge ($x = 1$) and Si ($x = 0$). Vertical arrows mark the respective indirect band gaps [Humlicek (1995)].

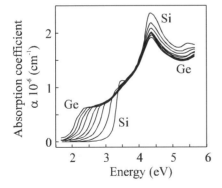

FIG. 6.4.5. Absorption coefficient α in the energy range between 1.5 and 5.5 eV for bulk Si$_{1-x}$Ge$_x$ alloys. Lines correspond (left to right) to composition values of $x = 1$ (Ge), 0.915, 0.831, 0.75, 0.635, 0.513, 0.389, 0.218, 0 (Si) [Humlicek et al. (1989)].

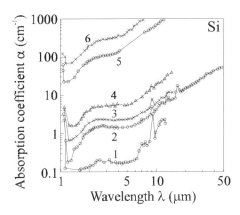

FIG. 6.4.6. Free electron absorption versus wavelength at different doping levels for n-Si ($x = 0$) at 300 K [Spitzer and Fan (1957)]. Electron concentrations (cm^{-3}): Curve 1, 1.4×10^{16}; curve 2, 2.8×10^{16}; curve 3, 1.7×10^{17}; curve 4, 3.2×10^{17}; curve 5, 6.1×10^{18}; curve 6, 1.0×10^{19} cm^{-3}.

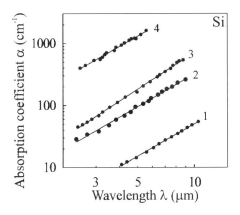

FIG. 6.4.7. Free hole absorption versus wavelength at different doping levels for p-Si ($x = 0$) at 300 K [Hara and Nishi (1966)]. Hole concentrations (cm^{-3}): Curve 1, 4.6×10^{17}; curve 2, 1.4×10^{18}; curve 3, 2.5×10^{18} cm^{-3}; curve 4, 1.68×10^{19}.

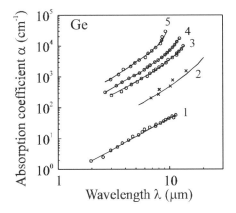

FIG. 6.4.8. Free electron absorption versus wavelength at different doping levels for n-Ge ($x = 1$) at 300 K [Fistul (1967)]. Electron concentrations (cm^{-3}): Curve 1, 8.0×10^{17}; curve 2, 4.8×10^{18}; curve 3, 1.35×10^{19}; curve 4, 1.8×10^{19}; curve 5, 3.6×10^{19}.

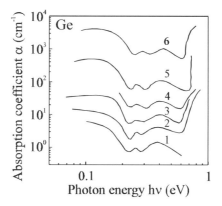

FIG. 6.4.9. Free hole absorption versus wavelength at different doping levels for p-Ge ($x = 1$) at 300 K [Ukhanov (1977) and Vasilyeva et al. (1967)]. Hole concentrations (cm^{-3}): Curve 1, 7.3×10^{15}; curve 2, 1.6×10^{16}; curve 3, 6.0×10^{16}; curve 4, 1.9×10^{17}; curve 5, 1.2×10^{18}; curve 6, 1.0×10^{19}.

The data on free carrier absorption for $Si_{1-x}Ge_x$ alloys at 78 K and 300 K can be found in Braunstein (1963).

6.5. THERMAL PROPERTIES

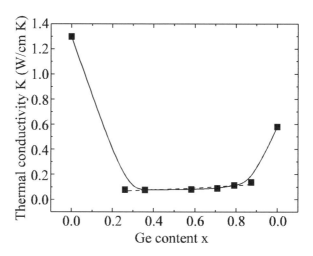

FIG. 6.5.1. Thermal conductivity of undoped $Si_{1-x}Ge_x$ alloys [Stöhr et al. (1954)].

At $0.25 < x < 0.85$

$$K = 0.046 + 0.084 \ (\mathrm{W/cmK})$$

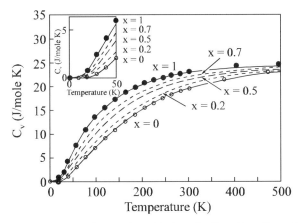

FIG. 6.5.2. Temperature dependence of specific heat in $Si_{1-x}Ge_x$ alloys. Inset shows the same dependences in the temperature range from 0 to 50 K [Wang and Zheng (1995)].

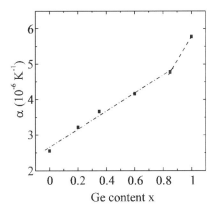

FIG. 6.5.3. Thermal expansion coefficient α at 300 K versus x [Zhdanova et al. (1967)].

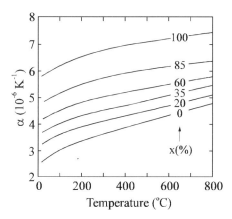

FIG. 6.5.4. Temperature dependence of the thermal expansion coefficient of Si$_{1-x}$Ge$_x$ alloys [Zhdanova et al. (1967)].

At $x < 0.85$

$$\alpha = (2.6 + 2.55x) \times 10^{-6} \ (\text{K}^{-1})$$

At $x > 0.85$

$$\alpha = (7.53x - 0.89) \times 10^{-6} \ (\text{K}^{-1})$$

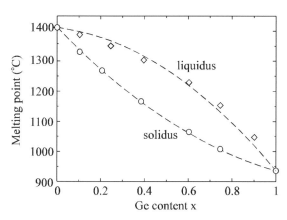

FIG. 6.5.5. Liquidus–solidus curves for binary Si$_{1-x}$Ge$_x$ alloys [Stöhr and Klemm (1954)]. **Liquidus:** $T_l \approx 1412 - 80x - 395x^2$ (°C). **Solidus:** $T_s \approx 1412 - 738x + 263x^2$ (°C).

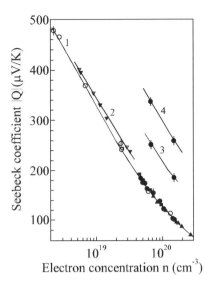

FIG. 6.5.6. Seebeck coefficient of n-type $Si_{1-x}Ge_x$ as a function of electron density at different temperatures. Closed symbols represent the data for the samples doped with P. Open symbols represent the data for the samples doped with As. Curve 1, 300 K. $x = 0.2$, 0.3, and 0.4. Curve 2, 300 K. $x = 0.8$. Curve 3, 600 K. $x = 0.15$. Curve 4, 900 K. $x = 0.15$ [Dismukes et al. (1964a)].

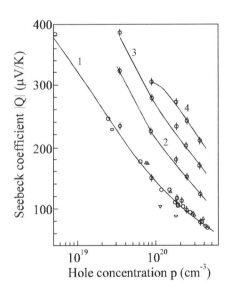

FIG. 6.5.7. Seebeck coefficient of p-type $Si_{1-x}Ge_x$ as a function of hole density at different temperatures. Curve 1, 300 K, $x = 0.2$, 0.3, and 0.4. Two symbols below curve 1 represent the data for $x = 0.7$ and $x = 0.8$. Curves $2 \div 4$, $x = 0.3$. Curve 2, 600 K; curve 3, 900 K; curve 4, 1200 K [Dismukes et al. (1964a)].

6.6. MECHANICAL PROPERTIES, ELASTIC CONSTANTS, LATTICE VIBRATIONS, OTHER PROPERTIES

Density: $(2.329 + 3.493x - 0.499x^2)$ g cm^{-3}

Surface microhardness
(using Knoop's pyramid test): $(1150 - 350x)$ kg mm^{-2}

Cleavage plane: {111}

Elastic constants at 300 K:

C_{11} $(165.8 - 37.3x)$ GPa
C_{12} $(63.9 - 15.6x)$ GPa
C_{44} $(79.6 - 12.8x)$ GPa

Bulk modulus B_s (compressibility^{-1}):

$$B_s = \frac{C_{11} + 2C_{12}}{3}, \qquad B_s = (97.9 - 22.8x) \text{ GPa}$$

Anisotropy factor

$$A = \frac{C_{11} - C_{12}}{2C_{44}}, \qquad A = (0.64 - 0.04x)$$

Shear modulus

$$C' = (C_{11} - C_{12})/2 \qquad C' = (51.0 - 10.85x) \text{ GPa}$$

[100] Young's modulus Y_0:

$$Y_0 = \frac{(C_{11} + 2C_{12})(C_{11} - C_{12})}{(C_{11} + C_{12})} \qquad Y_0 = (130.2 - 28.1x) \text{ GPa}$$

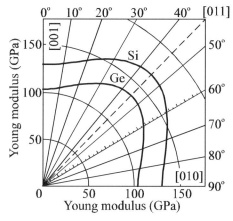

FIG. 6.6.1. Young's modulus as a function of direction in the (001) plane for Si and Ge [Wortman and Evans (1965)].

[100] Poisson ratio σ_0:

$$\sigma_0 = \frac{C_{12}}{C_{11} + C_{12}}, \qquad \sigma_0 = 0.278 - 0.005x$$

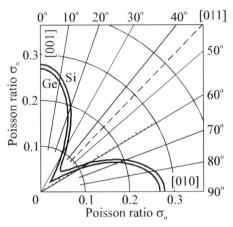

FIG. 6.6.2. Poisson ratio as a function of direction in the (001) plane for Si and Ge [Wortman and Evans (1965)].

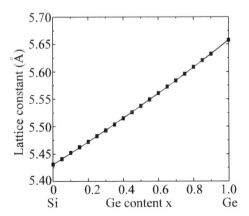

FIG. 6.6.3. Lattice constant **a** versus composition [Dismukes et al. (1964b)].

$$\mathbf{a}(x) = 5.431 + 0.20x + 0.027x^2 (\text{Å})$$

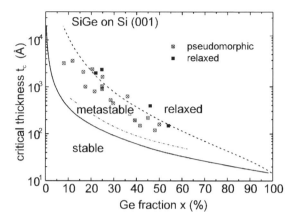

FIG. 6.6.4. The dependence of critical thickness for the onset of relaxation by the generation of misfit dislocations on Ge fraction x. Solid line represents equilibrium dependence. Dash–dotted curve represents the related dependence for the case when the strained Si$_{1-x}$Ge$_x$ film is capped by a Si layer. A fitting curve for a large number of experimental data referring to epitaxial growth at around 550 °C is shown as a dashed line [Herzog (1998)].

Dislocation Glide [Hull (1995)]

Dislocation glide velocity in bulk Si and Ge ($10 < \sigma < 100$ MPa)

$$v = a\sigma \exp(-E_v/kT) \ (\text{m/s}) \tag{6.6.1}$$

where

$a \approx 3 \times 10^{-3} \ \text{m}^2\text{s/kg}$
$E_v = 2.2 \ \text{eV}$ in Si ($x = 0$).
$E_v = 1.6 \ \text{eV}$ in Ge ($x = 1$).

To define the glide velocities in strained Si$_{1-x}$Ge$_x$ layers, normalized glide velocity v^* has been introduced:

$$v^* = \frac{v_m e^{-0.6x/kT}}{\sigma_{ex}} \tag{6.6.2}$$

Equation (6.6.2) normalizes the measured glide velocities v_m to an equivalent velocity at an excess stress σ_{ex} of 1 Pa in pure Si.

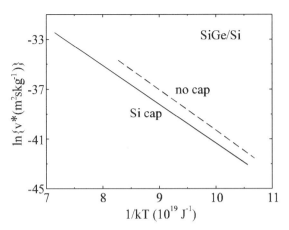

FIG. 6.6.5. Normalized dislocation velocities v^* for uncapped and capped (0.3 μm, Si) $Si_{1-x}Ge_x/Si(001)$ heterostructures versus $1/kT$ [Hull (1995)].

For uncapped SiGe:

$$v^* = \exp(-[7.8 + b/kT]) \; [m^2 s/kg] \qquad (6.6.3)$$

For Si-capped SiGe:

$$v^* = \exp(-[10.4 + b/kT]) \; [m^2 s/kg] \qquad (6.6.4)$$

where $b = 2$ eV.

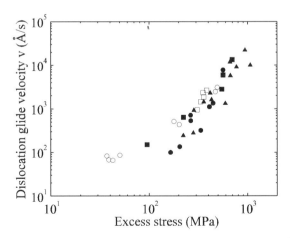

FIG. 6.6.6. Measured dislocation propagation velocities v_m in $Si_{1-x}Ge_x/Si(001)$ heterostructures versus excess stress at 550 °C [Hull (1995)].

Phonon Spectra

Si: Phonon Frequencies (in units of 10^{12} Hz)

LTO ($\Gamma_{25'}$)	15.5	TA (L_3)	3.45
TA (X_3)	4.5	LA ($L_{2'}$)	11.3
LAO (X_1)	12.3	LO (L_1)	12.6
TO (X_4)	13.9	TO ($L_{3'}$)	14.7

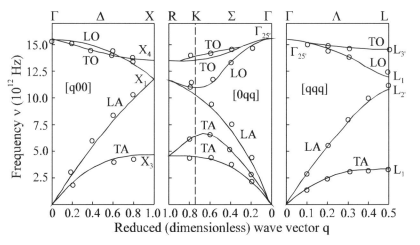

FIG. 6.6.7. Dispersion curves for acoustic and optical phonons in Si [Dolling (1963) and Tubino et al. (1972)].

Ge: Phonon Frequencies (in units of 10^{12} Hz)
[Nillsson and Nelin (1972)]

LTO ($\Gamma_{25'}$)	9.02	TA (L_3)	1.87
TA (X_3)	2.385	LA ($L_{2'}$)	6.63
LAO (X_1)	7.14	LO (L_1)	7.27
TO (X_4)	8.17	TO ($L_{3'}$)	8.55

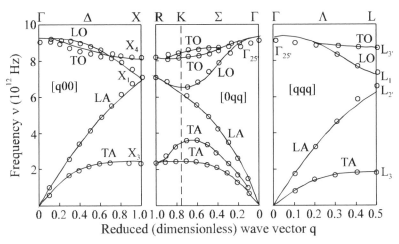

FIG. 6.6.8. Dispersion curves for acoustic and optical phonons in Ge [Weber (1977)].

For a wide composition range three optical phonon modes dominate the Raman spectra of $Si_{1-x}Ge_x$ alloys. They are attributed to local vibrations of Si–Si, Si–Ge, and Ge–Ge atom pairs. Their relative intensities are roughly proportional to the abundance of the respective pairs—that is, $(1-x)^2$, $2x(1-x)$, and x^2, respectively.

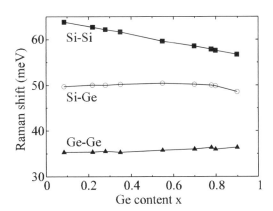

FIG. 6.6.9. Composition-dependence of the optical phonon Raman signals associated with local Si–Si, Si–Ge and Ge–Ge modes [Alonso and Winer (1989)].

REFERENCES

Alonso, M.I., and K. Winer, *Phys. Rev.* **B39**, 10056–10062 (1989).

Arafa, M., P. Fay, K. Ismail, J.O. Chu, B.S. Meyerson, and I. Adesida, *IEEE Electron Dev. Lett.* **17**, 124–126 (1996).

Balslev, I., *Phys. Rev.* **143**, 636 (1966).

Braunstein, R., A.R. Moore, and F. Herman, *Phys. Rev.* **109**, 695 (1958).

Braunstein, R., *Phys. Rev.* **130**, 869 (1963).

Busch, G., and O. Vogt, *Helv. Phys. Acta* **33**, 437 (1960).

Chandrasekhar, M., and F.H. Pollak, *Phys. Rev.* **B15**, 2127–2144 (1977).

Cheng, J.P., V.P. Kesan, D.A. Grutzmacher, T.O. Sedgwick, and J.A. Ott, *Appl. Phys. Lett.* **64**, 1681–1683 (1994).

Dismukes, J.P., L. Ekstrom, and R.J. Pfaff, *J. Phys. Chem.* **68**, 3021 (1964a).

Dismukes, J.P., L. Ekstrom, E.F. Steigmeier, I. Kudman, and D.S. Beers, *J. Appl. Phys.* **35**, 2899 (1964b).

Dolling, G., *Proceedings, Symposium on Inelastic Scattering Neutrons in Solids and Liquids*, IAEA, Vienna, **2**, 1963, p. 37.

Dutartre, D., G. Brémond, A. Souifi, and T. Benyattou, *Phys. Rev.* **B44**, 11525–11527 (1991).

Ebner, T., K. Thonke, R. Sauer, F. Schäffler, and H.-J. Herzog, *Phys. Rev.* **B57**, 15448 (1998).

Fistul, V.I., *Heavily Doped Semiconductors*, Nauka, Moscow, 1967 (in Russian).

Glicksman, M., *Phys. Rev.* **102**, 1496 (1956).

Hara, H., and Y. Nishi, *J. Phys. Soc. Japan* **21**, 1222 (1966).

Herzog, H.-J., unpublished data (1998).

Hull, R., in *Properties of Strained and Relaxed Silicon Germanium* (edited by Kasper, E.), EMIS Datareviews Series, No. 12, INSPEC, London, 1995, Chapter 1.3, pp. 28–45.

Humlicek, J., M. Garriga, M.I. Alonso, and M. Cardona, *J. Appl. Phys.* **65**, 2827–2832 (1989).

Humlicek, J., in *Properties of Strained and Relaxed Silicon Germanium* (edited by Kasper, K.), EMIS Datareviews Series, No. 12, INSPEC, London, 1995; Chapters 4.6 and 4.7, pp. 116–131.

Ismail, K., F.K. LeGoues, K.L. Saenger, M. Arafa, J.O. Chu, P.M. Mooney, and B.S. Meyerson, *Phys. Rev. Lett.* **73**, 3447–3450 (1994).

Kasper, E. (ed.), *Properties of Strained and Relaxed Silicon Germanium*, EMIS Datareview Series, No. 12, INSPEC, London, 1995.

Kline, J.S., F.H. Pollak, and M. Cardona, *Helv. Phys. Acta* **41**, 968–976 (1968).

Landoldt-Börnstein, *Numerical Data and Functional Relationships in Science and Technology*, New Series Group III, Vols. 17a and 22a, Springer, Berlin, 1982 and 1987.

Lang, D.V., R. People, J.C. Bean, and A.M. Sergant, *Appl. Phys. Lett.* **47**, 1333–1335 (1985).

Laude, L.D., F.H. Pollak, and M. Cardona, *Phys. Rev.* **B3**, 2623–2636 (1971).

Levinshtein, M.E., S.L. Rumyantsev, and M.S. Shur (eds.), *Handbook Series of Semiconductor Parameters*, Vol. 1: *Elementary Semiconductors and A_3B_5 Compounds, Si, Ge, C, GaAs, GaP, GaSb, InAs, InP, InSb*, World Science Publishing Co., Singapore, 1996.

Manku, T., and A. Nathan, *J. Appl. Phys.* **69**, 8414–8416 (1991).

Nelson, S.F., K. Ismail, J.O. Chu, and B.S. Meyerson, *Appl. Phys. Lett.* **63**, 367–369 (1993).

Nilson, G., and Nelin, *Phys. Rev.* **B6**, 3777–3786 (1972).

Penn, C., F. Schaffler, G. Bauer, and S. Glutsch, *Phys. Rev.* **B59**, 13314–13321 (1999).

People, R., *Phys. Rev.* **B32**, 1405–1408 (1985) and *Phys. Rev.* **B34**, 2508–2510 (1986).

Pickering, C., Carline, R.T., Robbins, D.J., Leong, W.Y., Barnett, S.J., Pitt, A.D., and Cullis, A.G., *J. Appl. Phys.* **73**, 239–250 (1993).

Rieger, M., and P. Vogl, *Phys. Rev.* **B48**, 14276–14287 (1993).

Schaffler, F., *Semicond. Sci. Technol.* **12**, 1515–1549 (1997).

Schmidt, E., *Phys. Status Solidi* **27**, 57 (1968).

Spitzer, W., and H.Y. Fan, *Phys. Rev.* **108**, 268 (1957).

Stern, F., and S.E. Laux, *Appl. Phys. Lett.* **61**, 1110–1112 (1992).

Stöhr, H., and W. Klemm, *Z. Anorg. Allgem. Chem.* **241**, 305 (1954).

Tubino, R., L. Piseri, and G. Zerbi, *J. Chem. Phys.* **56**, 1022–1039 (1972).

Ukhanov, Ju. I., *Optical Properties of Semiconductors*, "Nauka", 1977 (in Russian).

Van de Walle, C.G., and R.M. Martin, *Phys. Rev.* **B34**, 5621–5634 (1986).

Vasiljeva, M.A., L.E. Vorobyev, and V.I. Stafeev, *Fiz. i Tekhn. Polupr.* **1**, 1 (1967) 29–33 (in Russian).

Wang, K.L., and X. Zheng, in *Properties of Strained and Relaxed Silicon Germanium* (edited by Kasper, E.), *EMIS Datareview Series*, Vol. 12, INSPEC, London, 1995.

Weber, W., *Phys. Rev.* **B15**, 4789–47803 (1977).

Weber, J., and M.I. Alonso, *Phys. Rev.* **B40**, 5683–5693 (1989).

Whall, T.E., *J. Cryst. Growth* **157**, 353–361 (1995).

Wong, S.L., D. Kinder, R.J. Nicholas, T.E. Whall, and R.A. Kubiak, *Phys. Rev.* **B51**, 13499–13502 (1995).

Wortman, J.J., and R.A. Evans, *J. Appl. Phys.* **36**, 153 (1965).

Zhdanova, V.V., M.G. Kakna, and T.Z. Samadashvili, *Izv. Akad. Nauk. SSSR Neorg. Mater.* **3**, 1263 (1967).

Appendixes

APPENDIX 1. BASIC PHYSICAL CONSTANTS

Quantity	Symbol	Value
Avogadro's number	N_{AV}	6.02214×10^{23} mol
Bohr energy	E_B	13.6060 eV
Bohr radius	a_B	0.52917 Å
Boltzmann's constant	k	8.6174×10^{-5} eV/K
Electronic charge	q	1.6022×10^{-19} C
Electronvolt	eV	1.6022×10^{-19} J
Mass of electron at rest	m_0	0.91094×10^{-30} kg
Mass of proton at rest	M_p	1.6726×10^{-27} kg
Permeability in vacuum	μ_0	1.2623×10^{-8} H/cm $(4\pi \times 10^{-9})$
Permittivity in vacuum	ε_0	8.8542×10^{-12} F/m
Planck's constant	h	6.6261×10^{-34} J · s
Reduced Planck's constant	$\hbar = h/2\pi$	1.0546×10^{-34} J · s
Speed of light in vacuum	c	2.9979×10^{8} m/s
Wavelength of visible light	λ	0.4–0.7 μm

APPENDIX 2. PERIODIC TABLE OF THE ELEMENTS

Relative Atomic Mass Based on Carbon 12

4.00260
2 **He**
Helium

Atomic Number → Symbol, Name

VIII

IA	IIA	IIIB	IVB	VB	VIB	VIIB	VIII			IB	IIB	IIIA	IVA	VA	VIA	VIIA	
1.00794 1 **H** Hydrogen																	4.00260 2 **He** Helium
6.941 3 **Li** Lithium	9.01218 4 **Be** Beryllium											10.81 5 **B** Boron	12.0111 6 **C** Carbon	14.0067 7 **N** Nitrogen	15.9994 8 **O** Oxygen	18.9984 9 **F** Fluorine	20.179 10 **Ne** Neon
22.9898 11 **Na** Sodium	24.305 12 **Mg** Magnesium											26.9815 13 **Al** Aluminum	28.0855 14 **Si** Silicon	30.9738 15 **P** Phosphorus	32.06 16 **S** Sulfur	35.453 17 **Cl** Chlorine	39.948 18 **Ar** Argon
39.0983 19 **K** Potassium	40.08 20 **Ca** Calcium	44.9559 21 **Sc** Scandium	47.88 22 **Ti** Titanium	50.9415 23 **V** Vanadium	51.996 24 **Cr** Chromium	54.9380 25 **Mn** Manganese	55.847 26 **Fe** Iron	58.9332 27 **Co** Cobalt	58.69 28 **Ni** Nickel	63.546 29 **Cu** Copper	65.39 30 **Zn** Zinc	69.72 31 **Ga** Gallium	72.59 32 **Ge** Germanium	74.9216 33 **As** Arsenic	78.96 34 **Se** Seanium	79.904 35 **Br** Bromine	83.80 36 **Kr** Krypton
85.4678 37 **Rb** Rubidium	87.62 38 **Sr** Strontium	88.9059 39 **Y** Yttrium	91.224 40 **Zr** Zirconium	92.9064 41 **Nb** Niobium	95.94 42 **Mo** Molybdenum	(98) 43 **Tc** Technetium	101.07 44 **Ru** Ruthenium	102.906 45 **Rh** Rhodium	106.42 46 **Pd** Palladium	107.868 47 **Ag** Silver	112.41 48 **Cd** Cadmium	114.82 49 **In** Indium	118.71 50 **Sn** Tin	121.75 51 **Sb** Antimony	127.60 52 **Te** Tellurium	126.905 53 **I** Iodine	131.29 54 **Xe** Xenon
132.905 55 **Cs** Cesium	137.33 56 **Ba** Barium	**La-Lu** 57 71	178.49 72 **Hf** Hafnium	180.948 73 **Ta** Tantalum	183.85 74 **W** Tungsten	186.207 75 **Re** Rhenium	190.2 76 **Os** Osmium	192.22 77 **Ir** Iridium	195.08 78 **Pt** Platinum	196.967 79 **Au** Gold	200.59 80 **Hg** Mercury	204.383 81 **Tl** Thallium	207.2 82 **Pb** Lead	208.980 83 **Bi** Bismuth	(209) 84 **Po** Polonium	(210) 85 **At** Astatine	(222) 86 **Rn** Radon
(223) 87 **Fr** Francium	226.025 88 **Ra** Radium	**Ac-Lr** 89 103	(261) 104 **Unq** Unnilquadium	(262) 105 **Unp** Unnilpentium	(263) 106 **Unh** Unnilhexium	(262) 107 **Uns** Unnilseptium	(265) 108 **Uno** Unniloctium	(266) 109 **Une** Unnilennium									

Lanthanoid series

138.906 57 **La** Lanthanum	140.12 58 **Ce** Cerium	140.908 59 **Pr** Praseodymium	144.24 60 **Nd** Neodymium	(145) 61 **Pm** Promethium	150.36 62 **Sm** Samarium	151.96 63 **Eu** Europium	157.25 64 **Gd** Gadolinium	158.925 65 **Tb** Terbium	162.50 66 **Dy** Dysprosium	164.930 67 **Ho** Holmium	167.26 68 **Er** Erbium	168.934 69 **Tm** Thulium	173.04 70 **Yb** Ytterbium	174.967 71 **Lu** Lutetium

Actinoid series

227.028 89 **Ac** Actinium	232.038 90 **Th** Thorium	231.036 91 **Pa** Protactinium	238.029 92 **U** Uranium	237.048 93 **Np** Neptunium	(244) 94 **Pu** Plutonium	(243) 95 **Am** Americium	(247) 96 **Cm** Curium	(247) 97 **Bk** Berkelium	(251) 98 **Cf** Californium	(252) 99 **Es** Einsteinium	(257) 100 **Fm** Fermium	(258) 101 **Md** Mendelevium	(259) 102 **No** Nobelium	(260) 103 **Lr** Lawrencium

Masses in parentheses are masses of the most stable isotopes

APPENDIX 3. RECTANGULAR COORDINATES FOR HEXAGONAL CRYSTAL

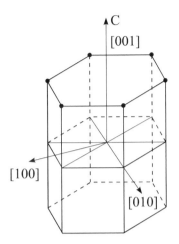

APPENDIX 4. THE FIRST BRILLOUIN ZONE FOR WURTZITE CRYSTAL

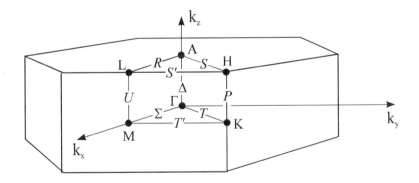

APPENDIX 5. ZINC BLENDE STRUCTURE

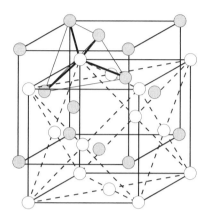

APPENDIX 6. THE FIRST BRILLOUIN ZONE FOR ZINC BLENDE CRYSTAL

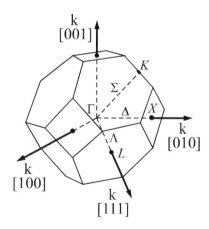

Additional References

Properties of GaN and related materials have been reviewed in the following books:

Edgar, J.H. (ed.). *Properties, Processing and Applications of Gallium Nitride and Related Semiconductors*, EMIS Datareviews Series, No. 23. INSPEC, London, 1999.

Morkoc, H. *Nitride Semiconductors and Devices*, Springer Series in Materials Science, Vol. 32, Springer, Berlin, 1999.

Nakamura, S. *Introduction to Nitride Semiconductor Blue Lasers and Light Emitting Diodes*, Taylor & Francis, New York, 1999.

Pankove, J.I., T.D. Moustakas, and R.K. Willardson (eds.). *Gallium Nitride (GaN) II. Semiconductors and Semimetals*, Vol. 57, Academic Press, San Diego, 1999.

Gil, B. (ed.). *Group III Nitride Semiconductor Compounds: Physics and Applications.* Series on Semiconductor Science and Technology, 6. Clarendon Press of the Oxford University Press, Oxford, England, 1998.

Moustakas, T.D., Jacques, I.P. (eds.). *Gallium Nitride (GaN) I. Semiconductors and Semimetals*, Vol. 50, Academic Press, San Diego, 1998.

Nakamura, S., and G. Fasol (eds.). *The Blue Laser Diode: GaN Based Light Emitters and Lasers*, Springer, Berlin, 1997.

Szweda, R. *Gallium Nitride & Related Wide Bandgap Materials & Devices: A Market & Technology Overview 1996–2001*, Elsevier Trends Division, Oxford, 1997.

The following books deal with the materials properties of SiC:

Choyke, W.J., H. Matsunami, and G. Pensl (eds.). *Silicon Carbide: A Review of Fundamental Questions and Applications to Current Device Technology*, 1st ed., Akademie Verlag, Berlin, 1997.

Harris, G.L. (ed.). *Properties of Silicon Carbide*, INSPEC, Institution of Electrical Engineers, London, 1995.

Dobson, M.M. *Silicon Carbide Alloys. Research Reports in Materials Science*, Vol. 11, Parthenon Press, Carnforth, Lancashire, England, 1986.

Properties of SiGe have been reviewed in the following books:

Hull, R., and J.C. Bean (eds.). *Germanium Silicon: Physics and Materials. Semiconductors and Semimetals*, Vol. 56, Academic Press, San Diego, 1999.

Yuan, J.S. *SiGe, GaAs, and InP Heterojunction Bipolar Transistors*, Wiley, New York, 1999.

Kasper, E. (ed.). *Properties of Strained and Relaxed Silicon Germanium*, EMIS Datareviews Series, No. 12, INSPEC, London, 1995.

Stoneham, A.M., and S.C. Jain (eds.). *GeSi Strained Layers and Their Applications*, Institute of Physics Publications, Philidelphia, 1995.

Jain, S.C. *Germanium–Silicon Strained Layers and Heterostructures. Advances in Electronics and Electron Physics*, Supplement 24, Academic Press, Boston, 1994.

MICHIGAN MOLECULAR INSTITUTE
1910 WEST ST. ANDREWS ROAD
MIDLAND, MICHIGAN 48640